Amazing
BUILDINGS

Illustrated by Paolo Donati and Studio Illibill

Written by Philip Wilkinson

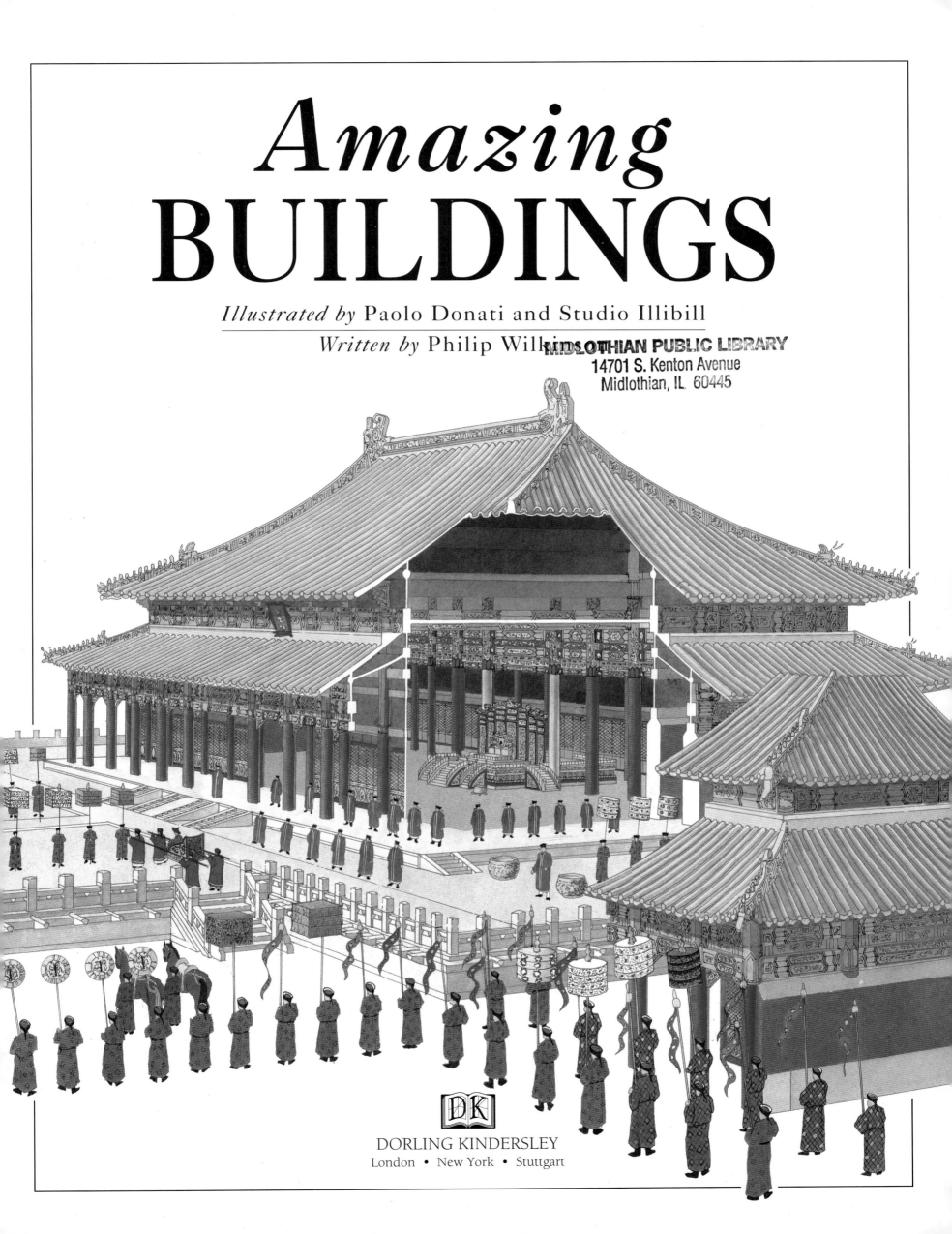

DK
DORLING KINDERSLEY
London • New York • Stuttgart

A DORLING KINDERSLEY BOOK

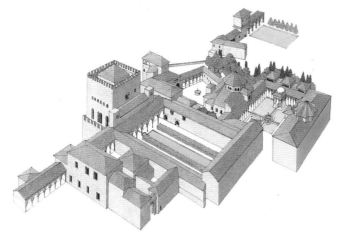

Art Editors Dorian Spencer Davies, Chris Scollen
Editors Emma Johnson, Anna Kunst
U.S. Editor B. Alison Weir
Architectural Consultant Andrew C. Smith

Managing Art Editor Jacquie Gulliver
Managing Editor Susan Peach

Production Shelagh Gibson, Marguerite Fenn

First American Edition, 1993
468109753
Published in the United States by
by Dorling Kindersley Inc., 232 Madison Avenue
New York, New York 10016

Library of Congress Cataloging-in-Publication Data
Wilkinson, Philip
Amazing buildings / by Philip Wilkinson;
Illustrated by Paolo Donati - 1st American ed.
p. cm.
Includes index
Summary: Provides the reader with glimpses inside twenty-one famous buildings from around the world.
ISBN 1-56458-234-5
1. Buildings - Juvenile literature [Buildings] I. Donati, Paolo, ill. II. Title.
NA2555.W48   1993   92-54314
720 - dc20   CIP
                                        AC

Reproduced by Dot Gradations, Essex
Printed and bound in Italy by New Interlitho, Milan.

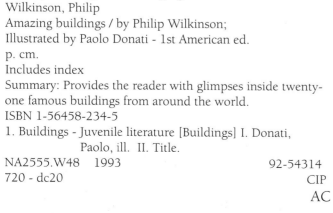

CONTENTS

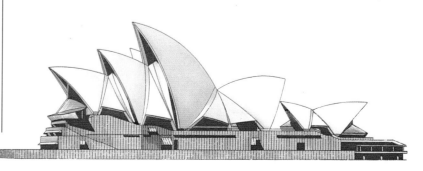

INTRODUCTION

OF THE MILLIONS OF BUILDINGS that there are in the world, some stand out because of their special qualities. These buildings may be larger, grander, or more beautifully decorated than the buildings that surround them, or it may be simply that they have been standing for many hundreds of years. Some are even more astonishing when you think of their size and complexity and the fact that they were built with nothing but muscle power and the most basic equipment. Even today, building a skyscraper calls for the manual skills of workers who climb over steel girders, high in the air, to bolt them in place. The stories behind the buildings are often as interesting as the structures themselves. They tell us about the people who built them and their lives. Why was a particular building built on a particular site? Why were certain materials used and what influenced the style of building? What is the significance of a building to the people who use it? Practical challenges, combined with social and cultural influences, are what make buildings so varied. A building may be built to withstand attack, to house a great art collection, to commemorate an event, or to glorify a god. Whatever the reason, buildings are an expression of the needs and hopes of people through the ages.

MESA VERDE
From the earliest times people used whatever materials they could find in their environment to make houses. Mesa Verde in Colorado, U.S.A., is a major site of Pueblo Indian cliff dwellings. The houses, which date from the 12th century, were made of stone and mud.

EARLY WOODEN BUILDINGS
Where trees were plentiful, people of the Stone Age made huts of wooden poles. The poles were joined together at the top to make an upside-down V shape. In the middle of the hut was an upright pole, which supported it.

STONE BUILDING
When the first metal tools came into use, stone could be shaped for building. Early stonecutters did a lot of the work at the quarry, carving huge blocks of stone to the required size and shape. This way they did not have to drag extra stone to the building site.

BRICKS AND MORTAR
Bricks were used where no other material was available, or where it was easier to transport. This was common in hot countries where bricks could be made of mud and dried in the sun. The bricks were laid with mud mortar in ways similar to modern bricklaying.

MASONRY
The Romans built roads and made use of rivers and canals to transport heavy materials to the building sites. They were skilled stone carvers, and many of their buildings have carved columns and other decorative features.

MODERN TIMBER FRAME
Wooden frames are still used for buildings today, particularly in countries with forests. Many buildings in the U.S. are based on a light timber frame, braced diagonally and fixed together with nails. Such buildings are quick, easy, and inexpensive to build.

STEAM POWER
By the late 19th century, cranes had been developed that could lift the large, heavy iron and steel pieces for framed buildings. These were powered by steam engines.

IRON AND STEEL
Many new opportunities in construction were opened up by the use of iron, and later steel, to make the frameworks for buildings. Tall, multistory buildings could be built and large rooms could be spanned without support columns.

CONCRETE, ANCIENT AND MODERN
The Romans were the first people to use concrete (made from a volcanic ash at that time). Today it is made with cement, and reinforced with steel rods. Concrete is a versatile and inexpensive material that is widely used in building.

TYPES OF STRUCTURES

In spite of their differences, most buildings have one of two basic structures. They are either framed buildings or walled buildings. In framed buildings, the weight of the building is carried on a frame of uprights and horizontals, and the walls can be light screens or be left out completely. In walled buildings, however, the weight is taken on solid platelike walls, which create rooms. Roofs, too, can be either solid, like domes, or based on a framework.

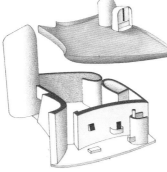

SOLID WALL

The heaviness of solid walls gives security and enclosure. The chapel at Ronchamp (pages 40-41) has solid stone walls. Brick or reinforced concrete can also be used for solid walls.

FRAMED WALL

The buildings in the Forbidden City, such as the Imperial Palace (pages 18-19), are supported on timber frames. Modern buildings often have steel or concrete frames.

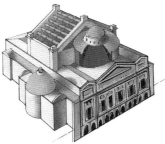

FRAMED ROOF

Most roofs are laid on a frame structure. An elaborate example of this is the Paris Opéra (pages 32-33). The frame can be made of either timber or steel.

SOLID ROOF

Domes, like that of the Taj Mahal (pages 24-25), are solid roofs. They can be made of brick, stone, or concrete.

1   2   3   4   5

6   7   8   9   10   11

THE MASTER MASON AND HIS TEAM

A Gothic cathedral, like most large buildings, was built by a large workforce, each member with special skills. The "masters" of each trade had a team working under them. The whole operation was supervised by the Master (stone) mason.

1 Master mason
2 Master quarryman
3 Stone setter (bricklayer)
4 Clerk of the works
5 Master stone carver
6 Master carpenter
7 Blacksmith
8 Plumber
9 Master glassmaker
10 Master roofer
11 Skilled laborer

IN CHARGE

The Master mason was both architect and chief builder. He had to know the art and science of design and building. Geometry was the practical science behind the cathedral. The Master mason would use this knowledge to make sure that the building was stable, and its proportions and detail were both pleasing to the eye and took into account the beliefs of the Church.

CARVING GARGOYLES

Inside and out, a Gothic cathedral was decorated with carvings. This work was supervised by the Master mason, but there was plenty of chance for the carvers to express themselves. Sometimes they produced gargoyles (grotesque figures) that were humorous rather than respectful.

BUILDING A CATHEDRAL

The great medieval cathedrals were huge, complex buildings that took many years to complete. The design often changed during the course of the work, as building problems cropped up, architectural fashions changed, or money ran out. This illustration shows builders working on a typical Gothic cathedral.

UP HIGH

Gothic cathedrals were the tallest buildings of their time. Wooden scaffolding and cranes with ropes and pulleys had to be built to haul materials up to the tops of high walls and towers.

BUILDING WINDOWS

The huge windows of a cathedral required special treatment. A wooden framework was built to support the heavy stones until the arch over the window was complete. The stonemason who laid the delicate stones of the window stood on a temporary platform.

THE PALACE OF MINOS

FOUR THOUSAND YEARS AGO, between 1900 and 1400 B.C., the Greek island of Crete was occupied by the Minoans. They were a civilized and artistic people who were ruled by a dynasty of kings, called Minos. In 1899 the English archaeologist, Arthur Evans, discovered the Minoan civilization when he excavated a site at Knossos in Crete. He found evidence of a vast palace or a temple complex built by the Minoans, which covered an area of 1,550 sq. yds (1,300 m²). The reconstruction of the building shown here is based on Evans's idea of what the palace of King Minos might have looked like. According to discoveries made at the site, the building consisted of rooms, corridors, and courtyards on many levels. This maze, together with the importance of the Minoan cult of the bull, may well have inspired the myth of the labyrinth and the Minotaur.

A view of what remains of the palace, parts of which were reconstructed by the archaeologist Arthur Evans at the turn of the century.

BULL'S HORN DECORATION
The horn-shaped decoration shows the importance of the bull in Minoan culture.

COLUMNS AND CONSTRUCTION
Much of the palace structure was made of wood and rubble. The columns that held up the floors were made from tree trunks that tapered (got thinner) at the base. This was a typical Minoan design.

STAIRCASE
When Sir Arthur Evans excavated the site at Knossos he found a mass of clay that he thought could be the remains of a staircase. This large staircase probably led to some of the main rooms on the first floor of the palace. He had the staircase rebuilt, even though there was little evidence to show what the original would have looked like.

FEMALE FASHION
Minoan women wore long topless dresses, sometimes with layered skirts. A figure of the Minoan snake goddess, found in a shrine at Knossos, wears this style of dress.

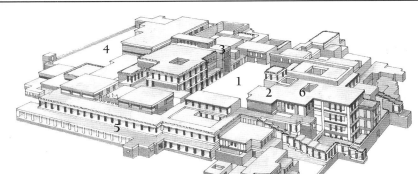

PLAN OF THE PALACE OF MINOS
1 Central court
2 Queen's rooms
3 Throne room
4 Storage rooms
5 South corridor
6 Hall of the Double Axes

TRADE CENTER
It is thought that the palace had several functions, as a center for collecting and distributing farm produce, as workshops for craftspeople, and as a depot from which goods were sent to the Aegean and eastern Mediterranean.

QUEEN'S ROOM
Among the private rooms of the palace is one thought to have belonged to the queen. It is decorated with paintings showing swimming dolphins. Dolphins were a symbol of freedom and happiness in the ancient world.

BULL-LEAPING IN THE CENTRAL COURT
The rooms were arranged around a courtyard in the center of the palace. This is where the famous bull-leaping ceremony, shown on wall paintings, may have taken place. Young acrobats would grab the horns of a bull and turn a somersault over its back. This ritual may have had some religious purpose. The young men and women who performed the "dance" were honored by the Minoans.

THE MINOTAUR
The bull was a symbol of strength and fertility to the Minoans, and the legend of the Minotaur is linked to the island of Crete. According to Greek mythology, the Minotaur was kept in a building called the labyrinth by King Minos, and was fed seven boys and seven girls from Athens every year in revenge for his son's death. Theseus, son of the king of Athens, killed the Minotaur with the help of Minos's daughter, Ariadne. She gave him a ball of thread so that he could find his way out of the mazelike labyrinth.

FRESCOES
The Minoans decorated their buildings with brightly colored wall paintings called frescoes. These pictures showed rituals that were important to the Minoans, for example bull-leaping and dancing, and creatures such as dolphins and fish.

COLOR AND PATTERN
The interior decoration was likely to have been friezes of spirals, rosettes, and other patterns.

THE COLOSSEUM

ONE OF THE FINEST BUILDINGS of ancient Rome still in existence, the Colosseum has one of the most bloodthirsty histories. In A.D. 72, the Roman emperor, Vespasian, ordered the Colosseum to be built as a stadium for spectacular events. He wanted to show off the soldierly qualities of courage and endurance so admired by the Romans, as well as provide entertainment for the people. With an incredible workforce of slaves and a system of movable scaffolding, it took about eight years to build the Colosseum. Regular events were held there, in which trained fighters, called gladiators, fought each other to the death in front of cheering crowds. Other spectacles included archery contests, boxing matches, swordswomen, chariot races, and wild-beast shows in which hundreds of animals were slaughtered in a single day. The last bloodthirsty spectacle staged in the Colosseum was in A.D. 523.

The Colosseum is nearly 2,000 years old and has survived several earthquakes. Today many people visit the ruins, which still stand in the center of Rome.

SUPERSTRUCTURE
The Colosseum is a huge oval structure. Its name comes from the Latin word *colosseus,* meaning "gigantic." It is 187 feet (57 m) high and measures almost 1,730 feet (527 m) around the outside walls.

THE ARENA FLOOR
The events took place on the arena floor, which was made of wood. Sand was spread on the floor to absorb any spilled blood. The word *arena* comes from the Latin word for sand or beach.

GLADIATORS
Gladiators were trained fighters. They were usually slaves and criminals who were trained in schools to fight each other to the death. Popular gladiators who survived were sometimes freed from slavery. Gladiators also fought wild animals, such as lions and tigers, which were imported from countries like Syria and north Africa. The gladiators used many weapons: nets and pronged spears (tridents), bows and arrows, and swords.

UNDERGROUND WARREN
Underneath the arena was a network of animal pens, rooms for the gladiators, cages for criminals, and ramps leading to different levels. Lifts hidden under trapdoors carried animals and people up to the arena floor.

STATUES
The outside of the building was decorated with statues of famous Romans and Roman gods.

UP AND OUT
Lifts operated
by pulleys
hoisted the
caged animals
through a
trapdoor to
the arena.

SUN CANOPY
An enormous canvas
canopy shaded spectators
from the sun. Very little
is known about how it
was put up or held in
place, but it must have
been an amazing feat of
engineering.

BY TICKET ONLY
People entered the
games with a ticket, just
like sporting events
today. Spectators sat on
tiers of seats behind the
emperor. First came the
Roman nobility and
wealthy citizens, then
slaves and foreigners,
and finally, at roof level,
the women.

ARCHES
The bottom three stories of the
Colosseum were arched to form a
gigantic passage with an arched
roof. These vaulted arcades are of
Roman concrete, brick, and stone.

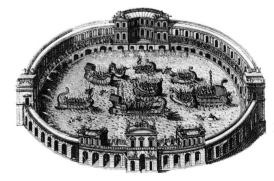

A NAVAL BATTLE
One of the most spectacular events ever staged at
the Colosseum was a mock sea battle. The floor
of the arena was removed, and the whole area
was flooded with water from a drained lake on
the site. Gladiators fought each other from ships
or in the water. Up to 100 ships and 19,000 men
took part in the pretend battle.

THE TEMPLE OF THE INSCRIPTIONS

DEEP IN THE RAIN FOREST of Central America stands a huge stone pyramid that is over 1,000 years old. The Temple of the Inscriptions was one of many temples built by the Maya, an ancient American Indian civilization whose most creative period was between A.D. 300 and A.D. 900. The Maya were brilliant mathematicians and astronomers, and they built temples such as this one to honor their gods of the sun, moon, and rain. One of the most important Maya cities was Palenque, in what is now southern Mexico. This is where the 75-foot (23 m) Temple of the Inscriptions was built, together with several other buildings. Hidden in the depths of the temple is a secret crypt containing the tomb of the Maya king, Pacal. He was a boy of 12 when he came to power, and he went on to become one of Palenque's greatest rulers. At some stage, probably at the end of the Maya period, the entrance to Pacal's tomb was sealed. This may have been to protect the king's body, and treasures buried with it, from vandals and robbers. The tomb was not disturbed until 1952, when it was discovered by Cuban archaeologist Alberto Ruz Lhuillier.

THE INSCRIPTIONS
One of the greatest achievements of the Maya was the development of a writing system. They used pictorial symbols called glyphs, like those on the Temple of the Inscriptions, in their calendar and astronomical calculations, and to commemorate the careers of their rulers.

Much of what was the Maya city of Palenque is now covered by the trees of the rain forest. Only the central area, with the palace and temples, is still visible.

EARTH BASE
A Maya pyramid is a huge platform made of earth or rubble, with an outer shell of stone. Its purpose was to provide a raised setting for the temple on top. The Temple of the Inscriptions also contains a tomb.

STEEP STONE STEPS
The nine stepped levels of the pyramid provided standing room for priests and other people taking part in large-scale religious ceremonies.

OFFERINGS TO THE GODS
The Maya people believed that the gods needed to drink blood to survive. They sacrificed animals such as deer, dogs, turkeys, and even fellow human beings, as gifts to their gods. Jaguars played a special part in Maya religion. The Maya saw many links between humans and jaguars, so, to them, sacrificing a jaguar was probably second in importance to a human sacrifice. Maya priests sometimes wore jaguar skins during these ceremonies.

THE MAYA
The Maya got together for religious ceremonies. The upper classes were people who held political or religious offices; the middle classes were scribes and craftspeople; the ordinary people were mainly farmers.

NEVER TO BE FORGOTTEN
The Temple of the Inscriptions gets its name from the three tablets of 620 glyphs just inside the entrance to the temple. These record the history of Palenque's rulers up to the time of Pacal, comparing him to the Maya gods, and his reign to events that happened years before.

PLAN OF THE TEMPLE OF THE INSCRIPTIONS

1 Temple of the Inscriptions
2 Palace complex
3 Temple of the Cross
4 Temple of the Sun
5 Temple XI

PRIESTLY PRIVILEGE
A great flight of steps leads to the temple at the top of the pyramid. Normally only the priests were allowed to go right to the top.

SECRET CRYPT
When the great ruler Pacal died, his body was put into a magnificently decorated tomb. This was the first tomb ever found in a Maya pyramid. The crypt in which it lay was covered with a huge stone slab, weighing 5 tons (tonnes). On the slab was a carved image of the king at the moment of death, falling toward the monster of the underworld.

PREPARING FOR THE JOURNEY
Pacal's body, wrapped in a shroud of cotton and sprinkled with a red dye, was decorated with jade objects (jade was a symbol of high status). A jade bead, used to pay for food in the underworld, was put in the king's mouth, and a jade mask was placed over his face. Food and water were left in the tomb for the journey to the underworld. During the burial ceremony the tomb was lit with torches.

11

KRAK DES CHEVALIERS

Banner of Knights Hospitalers

THE 11TH TO 13TH CENTURIES in Europe were the years of the Crusades (wars of the Cross) in the Holy Land. The Crusaders, or Christian armies, wanted to take control of the Christian holy places in Jerusalem from the Muslim rulers of the area, so that pilgrims from Europe could visit the Holy City again. But their aims went further than this. A group of Crusaders established a European kingdom in what is now Syria. This was partly to guard the route to Jerusalem, partly to make economic gains in the area. The kingdom had to be defended, something that was often beyond the means of one crusading lord. So the wealthy orders of knights, such as the French Hospitalers, took on the rebuilding of many of the castles. Of these, the most impressive was the Krak des Chevaliers (Castle of Knights), situated on a high point 25 miles (40 km) west of Homs in Syria. The French Crusaders took the castle in 1099 and probably began to rebuild it early in the 12th century. Krak had to be well defended and house a large garrison. But it was more than a fortress; it was also a community, especially when knights rallied from all over the surrounding area. This was the reason for the large number of rooms in the thick tower walls. At one time the castle even had a windmill on the north wall for grinding corn, and its own aqueduct and reservoir to provide drinking water for the troops as well as to fill the moats. The castle finally fell in 1271, to the Sultan Beibars, who tricked the Crusaders into surrender.

Commander's room

A view of Krak today shows the hilltop site and solid walls which make it such an impressive fortress. T. E. Lawrence (Lawrence of Arabia) thought it was "the best preserved and most admirable castle in the world."

Main reservoir

AQUEDUCT

The aqueduct was a form of bridge, built to channel water to the castle. It supplied the nine reservoirs inside the castle, which provided water for the troops as well as for horses and livestock. This special feature of the castle was also its weak point. If the enemy could destroy the aqueduct, they could cut off the water supply to the garrison.

KNIGHTS OF ST. JOHN

The Order of St. John of Jerusalem, founded in 1070, was also known as the Hospitalers because its original function was to care for the pilgrims in the Holy Land. Krak des Chevaliers was given to the Hospitalers in 1142. By the mid-12th century they were fulfilling a military role as well as that of a hospital.

DOUBLE WALLS

Krak is a concentric castle, which means that it is made up of two rings of buildings, one inside the other. The idea was that if attackers got through the outer wall, the knights would still be able to defend the inner buildings and push the enemy back from there.

COMMANDER'S ROOM

The castle's commander had a room in one of the inner towers. It was a circular, arched room with a large bay window and fine stone carving. This decoration dates from the mid-13th century, the same period as the main hall.

TOWERS

The height of the towers at Krak allowed those defending the castle to see for great distances in any direction. The rounded towers were designed to lessen the damage from missiles fired from seige engines. The three inner towers were the strongest in the castle.

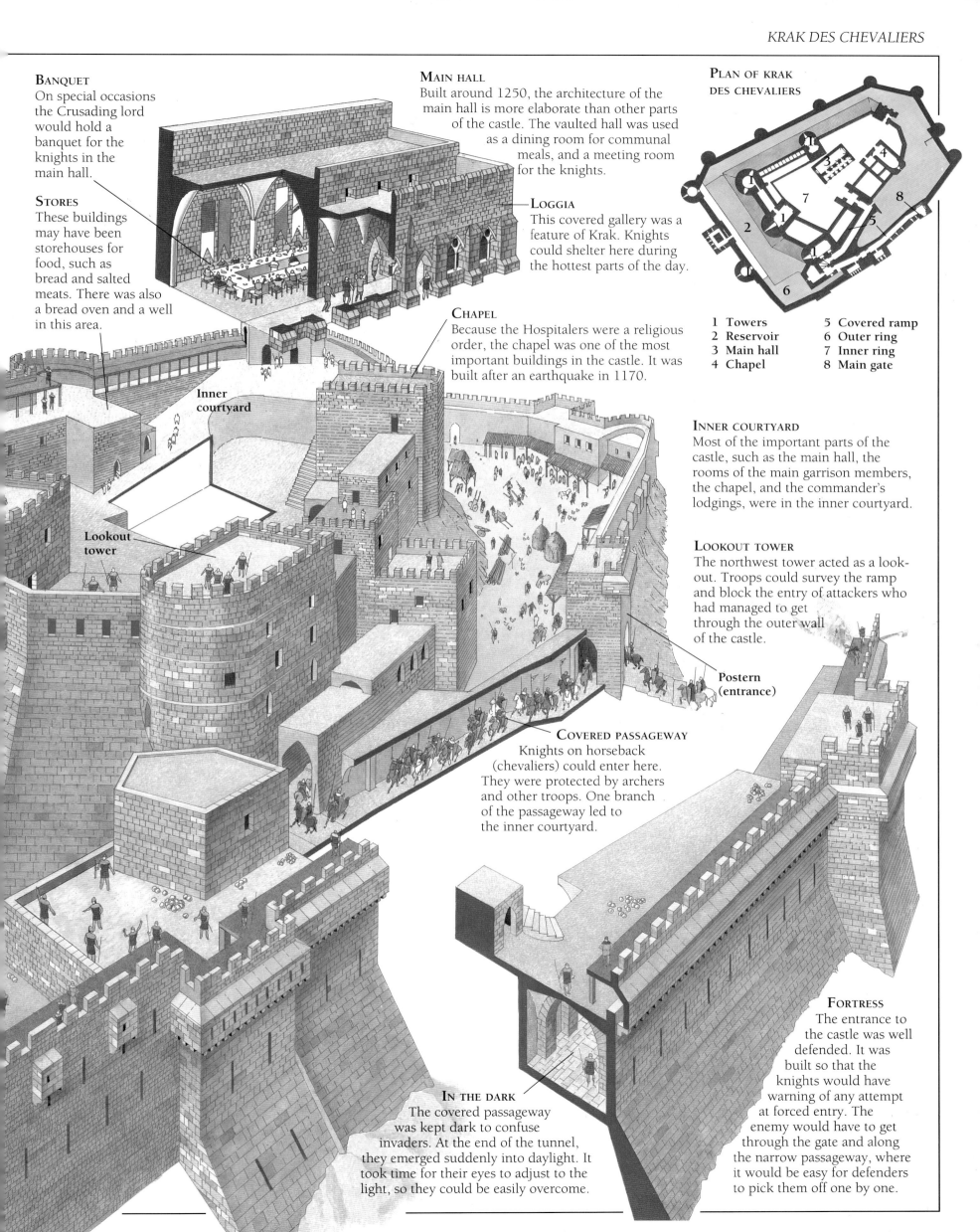

BANQUET
On special occasions the Crusading lord would hold a banquet for the knights in the main hall.

STORES
These buildings may have been storehouses for food, such as bread and salted meats. There was also a bread oven and a well in this area.

MAIN HALL
Built around 1250, the architecture of the main hall is more elaborate than other parts of the castle. The vaulted hall was used as a dining room for communal meals, and a meeting room for the knights.

LOGGIA
This covered gallery was a feature of Krak. Knights could shelter here during the hottest parts of the day.

PLAN OF KRAK DES CHEVALIERS

1 Towers
2 Reservoir
3 Main hall
4 Chapel
5 Covered ramp
6 Outer ring
7 Inner ring
8 Main gate

CHAPEL
Because the Hospitalers were a religious order, the chapel was one of the most important buildings in the castle. It was built after an earthquake in 1170.

Inner courtyard

Lookout tower

INNER COURTYARD
Most of the important parts of the castle, such as the main hall, the rooms of the main garrison members, the chapel, and the commander's lodgings, were in the inner courtyard.

LOOKOUT TOWER
The northwest tower acted as a lookout. Troops could survey the ramp and block the entry of attackers who had managed to get through the outer wall of the castle.

Postern (entrance)

COVERED PASSAGEWAY
Knights on horseback (chevaliers) could enter here. They were protected by archers and other troops. One branch of the passageway led to the inner courtyard.

IN THE DARK
The covered passageway was kept dark to confuse invaders. At the end of the tunnel, they emerged suddenly into daylight. It took time for their eyes to adjust to the light, so they could be easily overcome.

FORTRESS
The entrance to the castle was well defended. It was built so that the knights would have warning of any attempt at forced entry. The enemy would have to get through the gate and along the narrow passageway, where it would be easy for defenders to pick them off one by one.

13

THE ALHAMBRA

ENTERING THE ALHAMBRA IS LIKE entering the world of fairy tales. From the outside it looks like an ordinary castle, set in a hot, dry landscape, but behind the walls lie lush gardens, cool fountains, and shady courtyards. This palace-fortress in Granada, southern Spain, was built between 1238 and 1358 by Moorish princes, Muslims from North Africa who ruled Spain for several centuries. The Moors were a cultured people, and each ruler added some artistic detail to the building. One of the princes, Banu L' Ahmar, built a water course over one mile (1.6 km) long to supply the building and the fountains, pools, and channels that are such an important feature of the Alhambra. The name Alhambra comes from the Arabic word for "red," and describes the color of the sun-dried bricks of the outer walls. These great walls enclosed the three main areas of the palace – the Alcazaba (fortress), the Royal Palace, and a now ruined village for the courtiers and servants. The Moorish princes were driven out of Spain, in 1492, by the forces of Ferdinand of Aragon. They left behind one of the most beautiful palaces in the world and one of the last surviving monuments to a vanished civilization.

FORTIFIED TOWERS
The Alhambra had 24 towers and was big enough to contain an army of 40,000 soldiers within its walls.

The Alhambra's arched windows and colonnades are decorated with plaster, shaped in downward hanging forms like stalactites in a cave. Beneath the lacelike plaster work, the lower parts of the walls are covered with mosaic tiles in brilliant colors and dazzling patterns.

HALL OF TWO SISTERS
The hall is named in memory of two sisters who, the story goes, were imprisoned there and wasted away from lack of love. The sisters are represented by two stone slabs on either side of the fountain of lions. The Sisters' Hall is the best-preserved part of the Royal Palace and is thought to be one of the most beautiful.

INSPIRING BEAUTY
The Moors were skilled craftsmen who made patterns out of tiles, stone, wood, and plaster work. They used three types of decoration: plant design, calligraphy (Islamic inscription), and mathematical patterns. These beautiful designs were thought to inspire meditation.

MUSLIM PARADISE
As you walk through the courtyards of the palace you can hear the sound of running water, bubbling from fountains and springs. Most of the year, the smell of roses fills the air.

COURT OF THE LIONS
Named after the 12 stone lions at the base of the fountain, this is one of the main courtyards of the palace. It has arcades all around, and a pavilion at each end. On the rim of the 12-sided fountain bowl is a poem about the garden and its creator. The courtyard has a hidden meaning, too: a garden divided into four by water channels is an Islamic symbol of paradise.

From the outside, the Alhambra's great walls reveal nothing of the refined world within. Their size and strength make it clear how the palace survived as the last remaining fortress of the Muslims in Spain.

KEY TO PLAN OF THE ALHAMBRA
1 Hall of Abencerrajes
2 Court of the Lions
3 Hall of the Two Sisters
4 Gardens of Daraxa
5 Tower of the Ladies

FOREST OF COLUMNS

There are more than 100 columns in the Court of the Lions alone. With their delicately patterned arches, they create complex patterns of light and shade. This makes the large building seem smaller and more intimate. The arched colonnades also provide shade from the hot Spanish sun.

6 Court of the Myrtles
7 Hall of the Ambassadors
8 Mexuar (small palace)

HALL OF ABENCERRAJES

This finely decorated hall has a tragic story associated with it. One of the rulers of the Alhambra had all the Abencerrajes (the sons of his first wife) beheaded here. This was so that his son by his second wife would inherit the throne.

Court of the Lions

CHARTRES CATHEDRAL

A LONG PIECE OF SILK, thought to be the Virgin Mary's veil, was presented to the people of Chartres in France in the year 876. It was a gift from Charles the Bald, the grandson of the Emperor Charlemagne. This holy relic turned the town of Chartres, where an altar to the Virgin already existed, into one of the most important places of pilgrimage in the Western world. The veil survived a terrible fire in 1194, which destroyed most of the cathedral, and this was taken as a sign from the Virgin Mary that a new church must be built for her. Most of the new cathedral was built within 30 years – an amazing feat for medieval builders, who had no powered machinery to help them. It was finally completed in 1260. The aim of the master builder was to build the cathedral as high as possible, so that it seemed to reach toward heaven, and to include large stained-glass windows to flood the interior with colored light. Chartres is one of the finest cathedrals built in the Gothic style – a style based on pointed arches, which took hold in Europe from the 12th to the 16th century. The cathedral's beauty is due to its perfect proportions, brilliant stained-glass windows, and more than 2,000 magnificent carved statues.

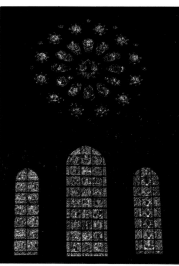

These stained-glass windows, which form part of the Royal Portal (doorway), show scenes from the Bible. The three lower windows are from the 12th century; the round rose window was added 100 years later.

The cathedral towers above the buildings around it – an effect that would have been even more noticeable when it was first built. Medieval pilgrims and visitors journeying to the cathedral would have seen the tall spires from far away.

FRAME ROOF
The original roof of Chartres Cathedral was supported by sloping wooden beams (rafters) and covered with lead. In 1836 an accidental fire destroyed the timbers, and they were replaced with a metal framework covered with copper sheeting.

SOARING VAULTS
The vaults (arched ceiling) of a cathedral were often built of stone. The ribbed cross vault was first used at Chartres. The ribbing linked the vault visually to the walls and arches below and made the weight of the stone easier to support. The cross vault allowed the windows to be built above the aisle roofs and flood the interior with light. This building method was copied and used in other cathedrals after Chartres, which had the highest vault of its day.

GLOWING COLORS
There are 176 stained-glass windows in the cathedral. The sunlight shining through the intense reds, blues, and violets of the colored glass creates a brilliant effect of light and color. Glass could only be made in quite small, irregular pieces in the 13th century, so the effect of the glaziers' work would not be seen until the windows were in place. The pieces of glass were assembled to make scenes from the Bible and the lives of the saints.

The central door of the Royal Portal, also known as the Portal of Glory, is surrounded by beautiful stone carvings of Christ and the saints.

SOUTH SPIRE

This spire, built in 1160, survived the fire of 1194 which destroyed most of the cathedral. On the first floor is an open, well-lit room, which was probably a workshop for the Master stonemasons. A fire in 1836 destroyed all the timbers of the upper level so there is no longer a belfry or bell.

NORTH SPIRE

In 1506, another fire, caused by a storm, destroyed the original north spire. The present spire was built between 1507 and 1513 and is 32 feet (10 m) taller than the south spire. Inside there are rows of bells that were rung to call people to worship, and a special room with a fireplace for the bellringers. The elaborate decoration of this spire contrasts with the simplicity of the earlier south spire.

TOWERING GOTHIC

Gothic cathedrals are full of images of "rising": the tall spires, the high interior vaults, the roofs reaching up to the sky. These are designed to make worshipers think of Christ rising to heaven, and to remind them of the presence of God above.

FLYING BUTTRESSES

The great flying buttresses (or supports) on the outside of the cathedral hold back the outward push of the vault. The stone buttresses also help transfer the force of the wind on the tall roof down to the ground. This is why they are known as "flying" buttresses.

PLAN OF CHARTRES CATHEDRAL

1 Royal (west) portal
2 North spire
3 South spire
4 Maze
5 Nave
6 North doorway
7 South doorway
8 Choir
9 Sacristy
10 Chapel of St. Piat

THE PIERS

The early cathedral builders used massive piers (pillars) to support the arches. However, this often made the interior seem dark because the piers shut out light from the windows. The builders of Chartres made the piers thinner to let more light in. They also wanted them to appear lighter in weight but still be strong enough to support heavy masonry. They did this by making the piers of solid stone, rather than filling them with rubble as builders had done in the past.

MAZE

Built into the floor of the nave at Chartres is a maze. Many medieval churches had one. The maze was probably a symbol of the path to Christ.

NAVE

This was the open area where people who came to worship stood. Today there is seating in the nave.

CRYPT

Many Gothic cathedrals have as much stone below ground as they do above. The stones below support and distribute the weight of the stones above. At Chartres the crypt houses important relics, such as the veil supposed to have been worn by the Virgin Mary. Pilgrims flocked in the thousands to look at the veil.

THE IMPERIAL PALACE

DURING THE PERIOD of the Ming dynasty in China (1368-1644), the Emperor Zhu Di (known as Yongle, or "Everlasting Happiness") ordered the old capital city to be rebuilt. He planned the new capital, Peking (now Beijing), in three parts. The outer, and largest, part contained houses, shops, and government buildings. Within this was the imperial city, an elegant area with lakes and gardens. And at the heart of the imperial city was the new palace, so large and with so many buildings that it seemed to be a city in its own right. Only the emperor's family, court, and people doing business with them were allowed into this central palace area. Ordinary people could not set foot inside the gates of the city, and so it became known as the "Forbidden City." The palace complex contains nearly 1,000 buildings – halls, temples, workshops, stables, and even a library. At its center are the Three Great Audience Halls: the Halls of Supreme Harmony, Complete Harmony, and Preserving Harmony.

At the front of the Forbidden City is the main northern gate, the Gate of Divine Military Genius, whose bell sounded 108 times every day at dusk. The southern gate, the Meridian Gate, was for the emperor only.

THRONE
Decorative coiled dragons form the back and armrests of the imperial throne, which is covered with gilding. The throne stands on a low podium, one of a series of platforms and steps leading from the square in front of the palace to the "peak" where the emperor sat.

THE EMPEROR
A special central area of the pavement was reserved for the emperor, who was carried in a palanquin, or covered litter.

HALL OF SUPREME HARMONY
This is the largest hall in the palace and the largest timber-framed building in China. It is 115 feet (35 m) high and raised on a marble podium. Twenty massive wooden columns support the huge roof. Here the emperor and his courtiers would gather for state celebrations, the New Year, and the emperor's birthday.

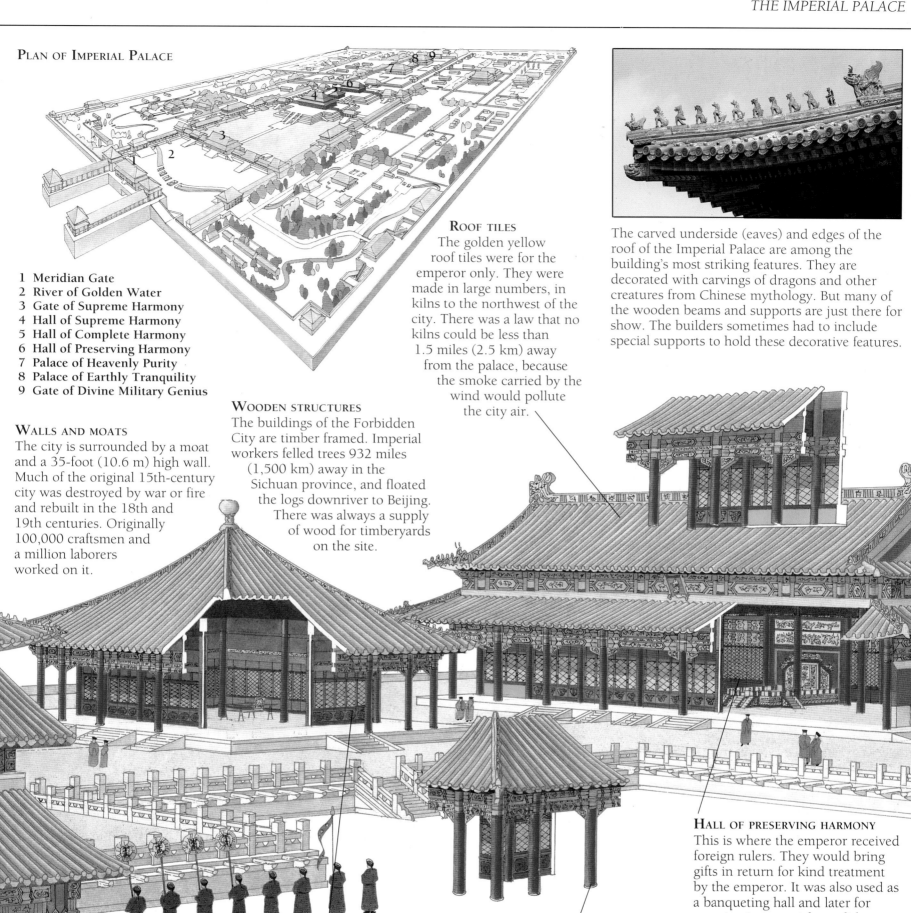

PLAN OF IMPERIAL PALACE

1 Meridian Gate
2 River of Golden Water
3 Gate of Supreme Harmony
4 Hall of Supreme Harmony
5 Hall of Complete Harmony
6 Hall of Preserving Harmony
7 Palace of Heavenly Purity
8 Palace of Earthly Tranquility
9 Gate of Divine Military Genius

WALLS AND MOATS

The city is surrounded by a moat and a 35-foot (10.6 m) high wall. Much of the original 15th-century city was destroyed by war or fire and rebuilt in the 18th and 19th centuries. Originally 100,000 craftsmen and a million laborers worked on it.

WOODEN STRUCTURES

The buildings of the Forbidden City are timber framed. Imperial workers felled trees 932 miles (1,500 km) away in the Sichuan province, and floated the logs downriver to Beijing. There was always a supply of wood for timberyards on the site.

ROOF TILES

The golden yellow roof tiles were for the emperor only. They were made in large numbers, in kilns to the northwest of the city. There was a law that no kilns could be less than 1.5 miles (2.5 km) away from the palace, because the smoke carried by the wind would pollute the city air.

The carved underside (eaves) and edges of the roof of the Imperial Palace are among the building's most striking features. They are decorated with carvings of dragons and other creatures from Chinese mythology. But many of the wooden beams and supports are just there for show. The builders sometimes had to include special supports to hold these decorative features.

HALL OF PRESERVING HARMONY

This is where the emperor received foreign rulers. They would bring gifts in return for kind treatment by the emperor. It was also used as a banqueting hall and later for examinations to pick candidates for the civil service. This hall does not have six golden pillars in front of the throne, as the others do, but the red furnishing adds to the atmosphere of splendor.

PILLAR AND POST

Additional rows of wooden pillars support the roofs of the verandahs around the halls.

HALL OF COMPLETE HARMONY

This small building was where the emperor put on his imperial robes for the great state occasions that took place in the Hall of Supreme Harmony. Here he also made announcements, and held meetings with his ministers and imperial guards. Sometimes he entertained and held banquets here. The hall is square in shape, unlike the halls on either side, which are rectangular.

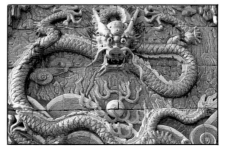

The dragon on this tile is an example of the dragon theme running through the Forbidden City. In ancient China, dragons were a symbol of goodness and strength. A dragon with five claws, holding a flaming pearl in waves or clouds, represents the emperor.

TEATRO OLIMPICO

IN THE CENTER OF THE NORTH Italian city of Vicenza is a 16th-century theater that became a model for all theaters and opera houses built thereafter. In 1580, when the Teatro Olimpico was built, Vicenza was the home of the *Accademia Olimpica* (Olympian Academy). This was a society of wealthy people and intellectuals whose particular interest was in the rediscovered plays of ancient Greece and Rome. We now call this rediscovery of the classical world the "Renaissance," a French word which means "rebirth." The Academy wanted their own theater where they could put on plays, and they wanted the design of it to be based on ancient Roman theaters. Andrea Palladio, who was a well-known architect of many buildings in and around Vicenza, and a founding member of the Academy, was chosen to design the new theater. He died during the first year of building, but work continued with another architect, Vicenzo Scamozzi, who designed the scenery and backdrops. The result was a theater unlike any built before, with seating copied from Roman theaters and stage scenery that was built in false perspective to look like a real street scene in the distance.

STATUES
Overlooking the stage and the auditorium are statues of the members of the Academy.

PAINTED SKY
Ancient Roman theaters were open-air structures with no roofs. In keeping with this style, the ceiling of the Teatro Olimpico was painted to look like the sky.

AUDITORIUM
The horseshoe-shaped auditorium (where the audience sat), and the seats arranged in tiers, are both copied from Roman theaters. Other classical features include a row of columns and statues, which enclose the auditorium. An eyewitness report of the theater on opening night told of its "incredible loveliness" because of the classical-style decoration.

THE AUDIENCE
Most of the spectators, who came from the upper classes of Vicenza society, entered through the back of the auditorium. They sat on tiered seats, which took up most of the area in front of the stage. The seats were not cushioned, as in theaters today. Spectators sat on hard benches for plays that often lasted several hours.

STAGE LIGHTING
The Teatro was one of the earliest indoor theaters (rather than open-air), so there was no natural light entering the auditorium. An early method of stage lighting was to place lamps behind rows of bottles filled with colored water.

The main wall of scenery, the *scenae frons,* is another feature from the Roman theater. It shows a grand architectural setting decorated with scenes from the life of the Greek hero, Herakles (Hercules), together with statues of members of the Academy.

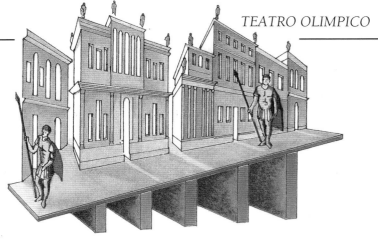

ABOVE THE STAGE
The ceiling over the stage was painted to look like the decorative ceiling of a grand room.

THE PERSPECTIVE SCENERY
Vicenzo Scamozzi took over the design of the Teatro after Palladio's death. He designed the perspective scenery, behind the *scenae frons*, to suggest an ideal city and act as a backdrop to any play. However, actors could not walk down the false streets because the streets made a full-size human look tiny!

THE IDEAL CITY
The street scene was said to represent the ancient Greek city of Thebes, but Scamozzi showed it as the ideal Renaissance city. This idea of a perfect city interested many Renaissance artists, and it can be seen in their paintings as well as their buildings.

THE STAGE
The stage has a setting of classical architecture (*scenae frons*) rediscovered from ancient Greece and Rome, as were the plays performed at the Teatro.

THE CAST
The actors' costumes had to be grand and extravagant, but they were not necessarily historically accurate. It was also important to have a variety of costumes, because although most Greek tragedies have only a few parts, each main character had a large group of attendants. The first performance at the Teatro was the Greek tragedy *Oedipus Rex*, by Sophocles. For this play, which has only nine characters, there was a cast of 108!

THE ORCHESTRA
This is the semicircular area in front of the stage where members of the audience sometimes sat.

PLAN OF THE TEATRO OLIMPICO

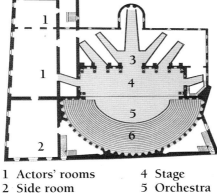

1 Actors' rooms	4 Stage
2 Side room	5 Orchestra
3 Perspective scenery	6 Seating

HIMEJI CASTLE

A FORTIFIED CASTLE SO MIGHTY that no enemy, however powerful, would be able to enter uninvited. This was the intention of the ruling warlords who built Himeji Castle, one of the largest and best-protected castles in Japan. The lords, known as *daimyo*, had large areas of land, and they were often at war with one another. Himeji Castle, built on the Harima plain in central Japan during the Middle Ages, started off as a small, three-story keep (fortified tower). By 1609 the warlord Ikeda Terumasa had added other buildings to it. He built walls, towers, moats, and other defenses, creating the castle complex that can be seen today. Inside a network of baileys (courtyards), protected by stone walls and 20 gatehouses, stands the great eight-story main keep and three linked smaller keeps. This central building was very difficult to attack. Invaders had to scale walls and pass through a series of twisting passages and narrow gateways before reaching it. Himeji was built at a time when firearms had just been invented, and the gunports – narrow openings through which archers and gunners could fire their weapons – were a special feature of the castle. Gunports and hidden openings allowed the defenders to drop rocks, or pour boiling water or oil, onto the enemy. From 1603 to 1868 there was a period of great peace in the country, and Himeji Castle changed from a fortress to a residence for the local *daimyo*. The castle's magnificence stems from a perfectly worked-out system of defense, combined with architectural elegance.

Himeji was nicknamed "White Heron Castle," because the white plaster and curved roofs reminded people of a wild bird seen in this part of Japan.

Japan is a country of wooden buildings, and the castles are no exception. These spacious, timber-structured rooms, typical of Himeji Castle, formed part of the living quarters of the ruling family.

MAIN COURTYARD
The main courtyard was at the center of the castle. Even if invaders penetrated this far into the defenses, they would still have to tread a winding path to get to the keep.

WARRIORS
Each country lord had his own castle and a band of warriors called *samurai*. These soldiers were highly skilled in the martial arts, and held a high position in society.

SIDE KEEPS
The small towers (*donjons*) on the corner of the main keep provided a good view of the courtyards below and of the approaching enemy.

DEFENSES
The strength of Japanese castles lay in the outer fortifications shielding the keep in the center. Further protection was given to the timber-framed keep with a covering of thick white plaster, which was fireproof as well as bulletproof. The courtyards around the castle's main keep were surrounded by thick, sloping stone ramparts. These walls were made from crudely shaped stones. For defense, they were built high as well as curved.

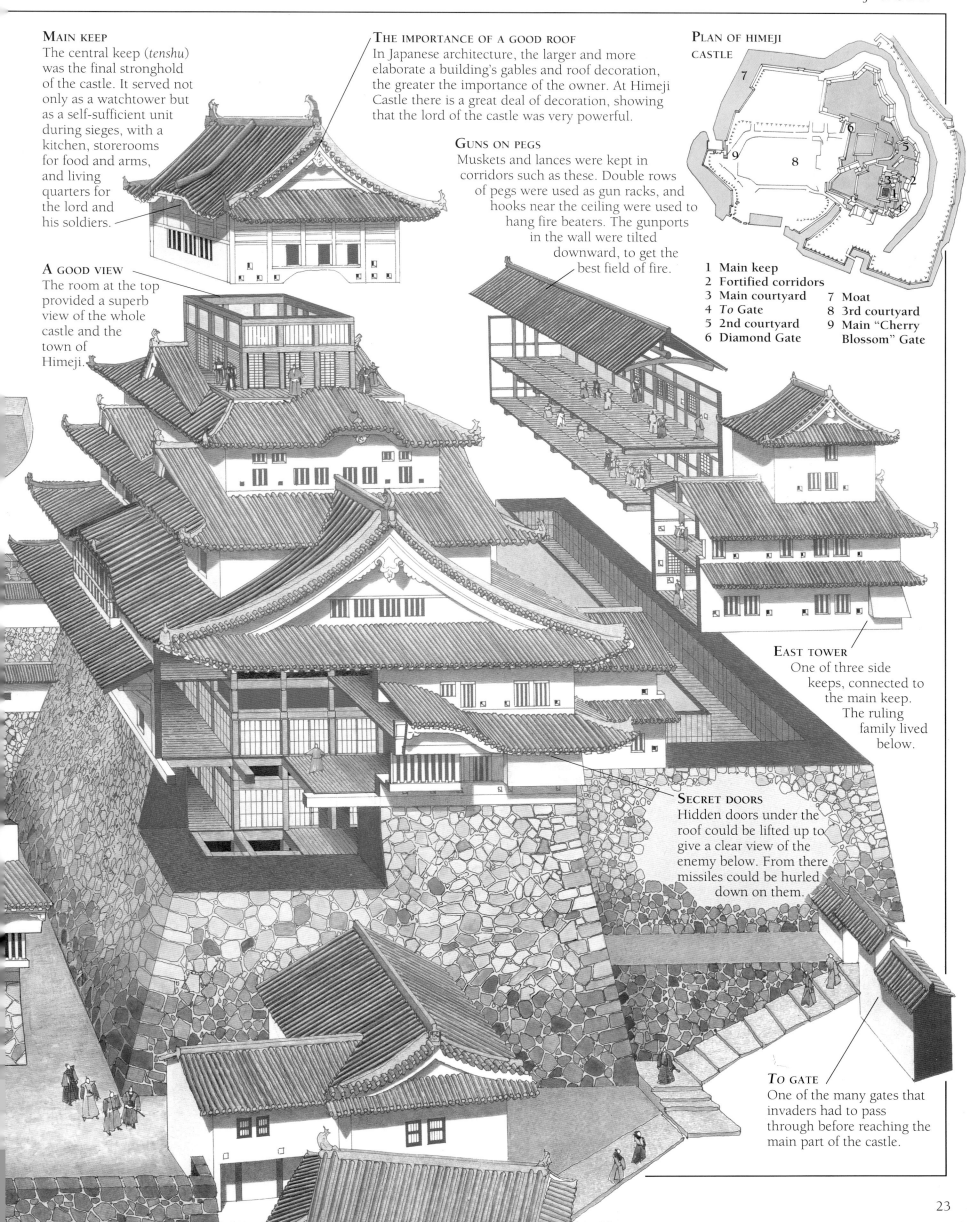

MAIN KEEP
The central keep (*tenshu*) was the final stronghold of the castle. It served not only as a watchtower but as a self-sufficient unit during sieges, with a kitchen, storerooms for food and arms, and living quarters for the lord and his soldiers.

THE IMPORTANCE OF A GOOD ROOF
In Japanese architecture, the larger and more elaborate a building's gables and roof decoration, the greater the importance of the owner. At Himeji Castle there is a great deal of decoration, showing that the lord of the castle was very powerful.

GUNS ON PEGS
Muskets and lances were kept in corridors such as these. Double rows of pegs were used as gun racks, and hooks near the ceiling were used to hang fire beaters. The gunports in the wall were tilted downward, to get the best field of fire.

PLAN OF HIMEJI CASTLE

1 Main keep
2 Fortified corridors
3 Main courtyard
4 *To* Gate
5 2nd courtyard
6 Diamond Gate
7 Moat
8 3rd courtyard
9 Main "Cherry Blossom" Gate

A GOOD VIEW
The room at the top provided a superb view of the whole castle and the town of Himeji.

EAST TOWER
One of three side keeps, connected to the main keep. The ruling family lived below.

SECRET DOORS
Hidden doors under the roof could be lifted up to give a clear view of the enemy below. From there missiles could be hurled down on them.

TO GATE
One of the many gates that invaders had to pass through before reaching the main part of the castle.

THE TAJ MAHAL

WHEN THE MOGUL EMPEROR'S beloved wife died in childbirth in 1631, he built her a tomb so beautiful that it was soon thought to be one of the most perfect buildings in the world. The name, Taj Mahal, is short for his wife's title, Mumtaz Mahal, meaning "Crown of the Palace." It took 20,000 workers 20 years to build the Taj Mahal, but the architect is unknown. Seen from afar, the tomb looks like a priceless jewel. The marble from which it is built had to be carried 310 miles (500 km) from Rajasthan to the bank of the Yamuna River in Agra, the Mogul capital of India. The Emperor, Shah Jahan, chose this spot not only for its beauty but also so that the building would be a place of refreshment and contemplation for the community, and a memorial to his wife. He could see the tomb from his palace and traveled to it along the river. Part of the perfection of the building lies in its symmetry, each half being a mirror image of the other. Shah Jahan was trying to produce an image of paradise on earth.

Jahan met Mumtaz, his stepmother's niece, at a New Year fair. He was 16 years old and she was 15. They married four years later. Jahan was allowed to have several wives, but Mumtaz was his favorite. She died giving birth to their fourteenth child.

The Taj Mahal stands perfectly reflected in the waters of a vast garden crisscrossed by canals. The water was collected from the river by a complicated system of buckets, tanks, and underground pipes, and fed into fountains. This style of garden is an Islamic vision of the Garden of Eden as paradise.

MINARETS
All mosques have towers called minarets, from which a crier calls people to prayer five times a day. The Taj Mahal was the first freestanding tomb to have minarets.

ON A PLINTH
The Taj Mahal stands on a marble base, called a plinth. This raises it above the garden level. The basement containing the real tombs of Mumtaz Mahal and Shah Jahan are therefore at ground level.

RAISING THE SKYLINE

The huge, marble dome is an impressive landmark on the skyline at Agra. It is 250 feet (76 m) high and is double domed. The height is gained by making the outside dome much higher than the inner ceiling. The space in between is hollow. The dome is a traditional Islamic symbol of womanhood and paradise.

A JEWEL-LIKE SURFACE

Shah Jahan loved gems and even designed his own jewelery. He had the Taj Mahal decorated in flower patterns using semiprecious stones, such as crystal and lapis lazuli. Flowers are very important in Islamic tradition as symbols of the divine kingdom.

NOT ALL MARBLE

The walls of the building are not solid marble, but layers of marble filled with rubble.

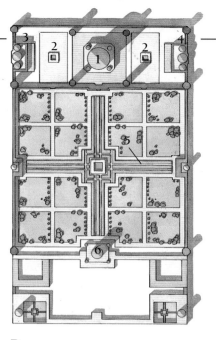

PLAN OF THE TAJ MAHAL
1 Royal tomb	4 Resthouse
2 Ponds	5 Canal
3 Mosque	6 Great gateway

A BUILDING TO READ

The Taj Mahal is covered with extracts from the Koran, the holy book of Islam. These texts are inscribed in beautiful handwriting. In the Islamic world, calligraphy – the art of handwriting – is the highest form of art. The Taj Mahal has more inscriptions than any other Mogul tomb.

SCREENED FROM VIEW

The delicately carved marble screen around the tombs took 10 years to build. It is more than 6 feet (2 m) high, and is decorated with flowers.

SIDE BY SIDE

The tombs that visitors see on the upper level are not the real tombs. These are hidden below, away from robbers and vandals. The tombs are beautifully decorated with 43 types of gems. Mumtaz's tomb is central while Jahan's is slightly higher and to one side. This is the only place where the symmetry of the Taj Mahal is broken.

VERSAILLES PALACE

ONE OF THE LARGEST PALACES in the world started life as a modest hunting lodge in the town of Versailles, just outside Paris. King Louis XIV of France (1638-1715) fell in love with the countryside around Versailles and transformed the lodge into a palace for himself. The kings and princes of 17th-century Europe wanted to show off their power, and the Palace of Versailles was Louis's way of doing this. Work started in 1660 and the palace took nearly 100 years to complete. The architect, Louis Le Vau, built separate suites of rooms for the king and queen around the original building. He was succeeded by Jules Hardouin Mansart, whose additions included the most magnificent room, the Hall of Mirrors, and two enormous wings to the north and south of the main block. At the same time, 250 acres (101 hectares) of gardens were laid out, with avenues, lakes, terraces, and sculptures. In 1682 Louis made Versailles the seat of French government and the entire court was moved there. The palace eventually became so large that people hired sedan chairs to carry them from one part to another!

The town of Versailles was planned around three broad avenues, which met just in front of the palace. The people of the town were constantly reminded of the presence of their king and his absolute power.

The magnificent gardens of Versailles were designed by a landscape artist called Le Nôtre. From every window of the palace, these formal gardens can be seen in all their glory.

AN IDEAL VIEW
From the terrace of the Hall of Mirrors the king and his courtiers could admire the gardens, and the park beyond.

HALL OF MIRRORS
This great hall, which was 235 feet (72 m) long, was the center for grand ceremonies and other public events. It was also the scene of many of Louis's lavish balls and dances.

STATUES AND MIRROR FRAMES
The Hall of Mirrors was completed in 1686. Some of the best artists in the country were employed to make the statues and frames.

THE SUN KING

Louis, who saw himself as Apollo (the ancient Greek god of the sun, music, and poetry), was called the "Sun King." The middle panel of the ceiling shows a picture of the Sun King at the height of his reign.

HALL OF LIGHT

It was common for the Sun King to appear before his courtiers in a room filled with light. So, while the windows let in natural light from the park, the chandeliers bathed the room in light from above. All this light was reflected in the full-length mirrors and magnified by them.

WALLS OF GLASS

The most striking thing about the design of the Hall of Mirrors is the way the mirrors on one wall match up with the windows on the garden side of the room. There are 17 of each.

MIRRORS IN PIECES

In the 17th century glassmakers could not produce single mirrors big enough for this room, so each mirror is made up of several panes of glass. Each pane measures about 38 x 28 in (89 x 71 cm).

THE KING'S BEDCHAMBER

The cutaway shows the king's bedchamber behind the Hall of Mirrors. From the windows there is a view down the avenue to Paris. This was a public as well as a private room. Courtiers attended the king's ceremonial *levée* – the audience he gave as he was getting up in the morning.

BRIGHTON PAVILION

GEORGE III WAS KING OF ENGLAND from 1760-1820. In his later years he suffered periods of illness, and during this time his son George ruled the country as Prince Regent. From time to time the Prince wanted to escape from the busy life of the court and to entertain his friends (and mistresses) in private. So, in 1783, he decided to rent a small farmhouse in the seaside town of Brighton. But he was not content with his modest house for long. In 1787 he asked the architect Henry Holland to enlarge it and to create a grander building in the classical style. This building was the Prince's summerhouse, or pavilion. Next the Prince discovered the Chinese style of decoration (according to some accounts when he was given a roll of Chinese wallpaper), and the Pavilion was redecorated inside in the Oriental style. This didn't happen until 1815, when he employed another architect, John Nash, to make further changes to the building. And what a transformation it was! The outside of the building included narrow towers called minarets, onion-shaped domes, and tentlike roofs. Inside, a magnificent banquet room and large kitchen were added. The result is a building that is truly amazing, a palace fit for a flamboyant prince.

THE ORIGINAL FARMHOUSE
Henry Holland's work on the Pavilion, in 1787, took in the Prince's original farmhouse (on the left-hand side of the new building). The old house had a pair of bay windows, which Holland changed to fashionable bow (curved bay) windows. He also added the large, circular, domed salon which had another bow-windowed block next to it.

DRAGONS AT DINNER
The Prince entertained in the banquet room, which was decorated using ideas from China. On the high domed ceiling is a plantain (banana) tree. Beneath it hangs a huge chandelier with winged dragon gas lamps, which were lit in the evening.

THE SALON
The main reception room, the salon, was used as a central hall for parties and meetings with the Prince.

Kitchen

Banquet room

A TON OF LIGHT

The dragon chandelier was the largest light of its time. It weighs one ton (1,000 kg).

ROOM FOR COOKING

The Prince was proud of his kitchen, which was probably the most up-to-date of its time. Twelve cooks worked there, tending steaming pans, hot plates, and large ovens. Meat was turned above the fire on a spit powered by the heat going up the chimney.

A MEAL TO REMEMBER

The dinner parties at the Pavilion were not the elaborate affairs of some of the European royal houses, when hundreds of people attended. Usually 30 or 40 people sat around the Prince's table. But the meals they ate were enormous. One surviving menu from the time lists 100 different dishes.

KITCHEN COLUMNS

The iron and copper palm-tree columns are one of the most striking features of the kitchen. They are functional, providing support for the heavy roof above. They also show Nash's skill in using new materials in imaginative ways in his buildings.

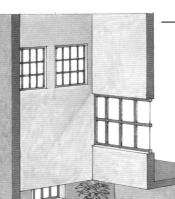

MUSIC ROOM

The Prince Regent, who played the cello, insisted on a music room for the Royal Pavilion. It was fitted with a huge organ, the biggest in England at the time of its construction. Many famous musicians played there, including the composer Rossini.

Nash based his design on Holland's original, but succeeded in turning a simple pavilion into an Oriental palace. Much of the decoration was built onto the original structure. Nash held up the domes and tent roofs with supports made from cast iron – a material quite new to architecture at that time.

TENT ROOFS

Nash created tent-shaped roofs for the banquet and music rooms. This was to vary the roofs on the skyline.

APPLIED DECORATION

Much of the wall decoration in the Pavilion is not very firmly attached to the walls. Flat wooden strips (battens) fixed to the brickwork were used to support the decorated panels. These formed a backing for silk or paper wallcoverings, gilding, and paint.

Salon

Music room

THE HOUSES OF PARLIAMENT

TRAGEDY STRUCK THE ANCIENT PALACE of Westminster in London on the night of October 16, 1834. A stove overheated and caused a fire so fierce that it almost destroyed the whole building. Very little, apart from Westminster Hall, was left standing. Originally a royal residence, the Palace of Westminster had been the headquarters of the British government for centuries. A new palace had to be built as soon as possible, and it had to be larger than the original to accommodate a government that had grown over the years. Sir Charles Barry, who was a well-known architect of the time, was chosen to design the new building. He worked with a young architect called Augustus Pugin. Barry designed courtyards that let natural light and fresh air into the heart of the building, and arranged the rooms so that it was easy to walk freely from one to the other. Pugin was responsible for the smaller details, designing everything from ornaments to hatstands and inkwells. Barry's well-planned scheme included rooms for the king or queen, the lords, and the commoners, and a network of committee rooms, offices, libraries, and lobbies. There are 100 staircases and more than two miles (3 km) of corridors. The building was completed in 1870, by which time both Barry and Pugin were dead. To this day, certain traditional ceremonies still take place in Parliament. At the State Opening of Parliament, the Queen's Messenger, Black Rod, calls the Members of the House of Commons to the House of Lords by knocking on the door with his rod, or mace.

The Houses of Parliament are situated on the north bank of the River Thames in the center of London.

YES OR NO?
To vote at the end of a debate, MPs (Members of Parliament) go into one of two lobbies – the "aye" (yes) lobby or the "no" lobby, according to whether they agree or disagree with the argument put forward. Officials count their votes as they pass, one by one and bowing, through the door.

CENTRAL LOBBY
Visitors who come to watch a debate must go first to the Central Lobby. This is the crossroads of the building. People who want to lobby (influence the opinions of) their MP gather here also.

Westminster Hall

WESTMINSTER HALL
Built at the end of the 14th century, this is the largest surviving part of the ancient Palace of Westminster. It is famous for its wooden hammerbeam roof, which is 70 feet (21 m) wide, a considerable feat of carpentry at the time. Many important state ceremonies took place in the hall, and law courts often sat here.

VICTORIA TOWER
This tower contains more than two million documents. These include the death warrant of Charles I, King of England, and copies of every Act of Parliament passed since 1497.

St. Stephen's Hall

Central Lobby

Lords' Lobby

HOUSE OF LORDS
The most lavish decoration is in the House of Lords, the debating chamber of nonelected members of Britain's "Upper House." The Lords sit on red leather benches (the color of all the upholstery in the Lords' rooms). Richest of all is the throne, used by the monarch at the State Opening of Parliament.

Prince's Chamber

Royal Gallery

Robing (dressing) Room

HOUSE OF COMMONS

This is where the elected Members of Parliament meet to discuss their ideas. The House is arranged with the government on one side and the opposition parties on the other. The two sides are further separated by a rule that says no member can cross the red line on the carpet in front of their own front benches. The lines were placed two swords' lengths apart so that members could not fight a duel! The Commons was damaged by fire during World War II. It was later restored following the original design and keeping the green color for benches and carpets. The House has 437 seats, but 650 MPs.

CLOCK TOWER

The clock tower, better known as Big Ben (the name of the bell inside), is probably the most famous part of the whole building. It is 316 feet (96 m) high, and has 334 steps leading up to the belfry. The tower was built from the inside outward, so no scaffolding was needed and the building seemed to grow as if by magic. Inside there is a cell, intended for members who have been rude or disruptive during debates. This was last used in 1880.

PRESS GALLERIES

These galleries are for journalists. The Strangers' Gallery at the other end is for members of the public who want to watch debates.

ORDER!

The Speaker is the chairperson, who keeps order during debates.

AT WORK

Since 1885, a lamp has been lit in the lantern at the top of the tower whenever Parliament sits at night.

IN THE BELFRY

Big Ben, the great bell that strikes the hours, is one of London's famous landmarks. It was probably named after Sir Benjamin Hall, the first Commissioner of Works, whose name is carved on it. There are four other bells in the belfry, but Big Ben is the biggest, weighing more than 13 tons (13.2 tonnes). In 1859 a crack in the bell silenced it for three years. The cure was to turn the bell so that the hammer struck a different spot. It has been ringing on time ever since.

THE CLOCK

Designed to be both accurate and clearly seen from a distance, the clock has four faces, each measuring 23 feet (7 m) across. Each Roman numeral is 2 feet (60 cm) high. Galleries behind the clock faces contain lights that illuminate the clock at night.

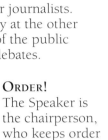

House of Commons

Commons Inner Court

Lords' Library

Commons Library

Terrace

THE PARIS OPÉRA

THE FRENCH EMPEROR NAPOLEON III had a passion for buildings and left many grand monuments behind. Among other projects, he was responsible for the replanning of the city of Paris, with broad new boulevards (streets) sweeping across the narrow streets of the old capital. At the junction of some of these new roads, a splendid new opera house was built in the mid 1800s. The architect, whose design was chosen from over 150, was Charles Garnier. The Opéra, which was started in 1862, took 13 years to complete, and was designed to make going to the theater a spectacular event. To Napoleon, it was a building that symbolized the triumph of his Empire. The Opéra was modeled on the architecture of Italian palaces and is an ideal setting for both the grand entertainments that took place on stage, and the wealthy and cultured audiences. Garnier supervised the project down to the last detail; few other buildings have had such a personal creative input by the architect. Today the Opéra is still one of the most luxurious theaters in the world, thought by many people to be a masterpiece of 19th century architecture.

THE CENTRAL DOME
Garnier exaggerated the dome above the auditorium, raising it 42 feet (13 m) higher than was necessary. This was to prevent the flytower (where the machinery for the scenery is housed) from dominating the view of the building from the front. The dome was built with a hidden iron framework. A strong mechanism was built to hold the massive chandelier that hangs above the auditorium.

BALCONIES
Garnier surrounded his great staircase with balconies. From there the rich could admire, and be admired by, their friends and acquaintances coming up the stairs. This was a popular French tradition and one that made the audience feel as if they were on stage themselves.

FIRST FLOOR
At the top of the stairs, where the staircase divides, is the entrance to the main part of the theater. The grand foyer beyond is a richly decorated room with columns, statues, mirrors, chandeliers, and luxurious drapery.

GRAND STAIRCASE
Beyond the ground floor entrance hall is the grand staircase, one of the most striking features of the Opéra. Its wide marble steps and Italian-style decoration provided a perfect backdrop for the richly dressed operagoers as they went up to their seats. Marble, onyx, and bronze cover the metal structure of the stairs.

GUARDED BY APOLLO

Garnier made the Paris Opéra a celebration of music and dance. On top of the flytower is a statue of Apollo, the Greek god who, among other things, took an interest in art and music. The other statues that decorate the outside of the Opéra include the Muses of Greek mythology, goddesses associated with singing, music, dance, drama, and poetry.

THE FLYTOWER

One of the most fascinating aspects of opera, theater, or ballet is the stage scenery, which may range from painted scenes of towering forests to the inside of a fairy-tale castle. These rich backgrounds are changed during the performance and are lowered from the area behind the stage, called the flies. Garnier designed a vast flytower for the huge wheels and pulleys that are used to lift or lower the flats (scenery).

The Paris Opéra stands at the intersection of several streets, an imposing landmark that is impossible to miss. From the front the three main sections of the interior (the salon, the auditorium, and the flytower) are reflected in the building's outside appearance. The large first-floor windows indicate the entrance salon and staircase beyond; the green dome covers the auditorium, and the flytower, with its triangular roof, is clearly visible beyond.

BEHIND THE SCENES

The rooms at the back of the building include offices, rehearsal rooms for dancers, singers, and actors, and changing rooms for the performers.

SCENERY AND PROPS

Garnier planned the Opéra to be self-sufficient. The scenery was made in workshops nearby. Garnier provided sewing and tapestry workshops, and an accessory department at the back of the building for other props (including costumes, hats, and jewelry). The facilities also included police and fire services, in case of an emergency!

ON STAGE

The stage at the Opéra is vast – it can hold up to 450 performers comfortably. This allows the company to stage grand opera and large-scale ballets and musicals.

PLAN OF THE PARIS OPÉRA

THE AUDITORIUM

The part of the theater where the audience sits is horseshoe shaped. It has four tiers of boxes, and seats just over 2,000 people. The auditorium is small in relation to the rest of the building, but the Opéra was still one of the largest opera houses of the time. The first opera to be staged in the new building was Halévy's *La Juive*, in 1875.

SPECIAL HALL

Beneath the auditorium, Garnier built a special entrance hall for the most privileged members of the audience – the wealthy people who had paid to have the Opéra built and who held season tickets. They could enter the Opéra straight from their carriages, which drew up alongside the covered entranceway. They would leave their cloaks there and, in their fine opera clothes, join the rest of the audience upstairs.

1 Main entrance
2 Main foyer
3 Grand staircase
4 Auditorium
5 Stage
6 Dancers' rooms
7 Offices
8 Emperor's entrance
9 Entrance for season-ticket holders

NEUSCHWANSTEIN CASTLE

PERCHED HIGH ON A ROCKY CRAG in southern Germany is Neuschwanstein Castle, one of the most fantastic buildings ever dreamed up. It was the creation of King Ludwig II of Bavaria, who ruled his country, reluctantly, from 1864 to 1886. Ludwig was the patron of the composer Richard Wagner and his operas. He made Wagner's operas, and the German legends that inspired them, the theme for his castle.

Neuschwanstein took 17 years to build. It was designed by a Munich scene painter, and built to look like a medieval castle, with fabulous towers, turrets, and battlements. However, it was much more luxurious than a medieval castle would have been. A central heating system was installed, and the kitchen had hot and cold running water. No expense was spared, but sadly Ludwig's "dream" castle was never finished. In 1886 he drowned in a nearby lake, after being declared unfit to rule. Parts of the building were never completed, including the room that was to house his sumptuous throne.

Ludwig II spent most of his time and money traveling around his kingdom, supporting Wagner's new operas and overseeing building projects. He particularly liked to go on midnight sleigh rides.

MAIN TOWER
Inside this tower is the main stair-case, leading to the castle keep.

SEAT OF GOLD
Ludwig II planned to put his lavish throne of gold and ivory on this spot.

THRONE ROOM
This large two-storied hall is decorated in rich colors, particularly gold, and has a patterned mosaic floor. Both of these features are typical of the Byzantine style, developed in the Eastern Roman Empire of Byzantium, now Istanbul. Ludwig II became fond of this style late in his life.

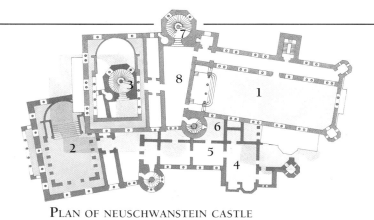

PLAN OF NEUSCHWANSTEIN CASTLE

1	Singers' Hall	5	Dining room
2	Throne Hall	6	Study
3	Domed ceiling	7	Round tower
4	King's bedroom	8	Lounge

MEDIEVAL STYLE
The circular towers give the skyline of Neuschwanstein its romantic, fairy-tale quality. The battlements and parapets are based on those of authentic medieval castles. At Neuschwanstein, these features were purely decorative; they had no practical use for defending the castle.

SINGERS' HALL
This room is partly modeled on a similar hall in the ancient German castle of the Wartburg, near Eisenach, where a famous medieval singing competition was held. The room also shows the influence of Wagner's opera *Tannhäuser,* which is set in and around the Wartburg. The decoration includes the symbols of the signs of the zodiac, and pictures of characters from the Parsifal saga, the basis of another of Wagner's operas.

Neuschwanstein Castle is stunningly situated on top of a crag overlooking the Alps and their foothills. It resembles the romantic castles of Wagner's operas.

PRIVATE ROOMS
Below the Singers' Hall are the king's private rooms – his study, dining room, and the bedroom, which is entirely devoted to the Wagner opera *Tristan and Isolde.* His huge bed, richly carved in oak wood, took 17 men over four years to complete.

Singers' Hall

KNIGHTS' HOUSE
The square tower is connected to the main rooms (the keep) by this building, which provides extra accommodation.

SQUARE TOWER
The square tower, which rises from the northern wall of the castle, is 148 feet (45 m) high. From the balcony there are dramatic views over the rivers, lakes, and valleys of the surrounding countryside.

KEMENATE
The Kemenate is the three-storied building in the upper courtyard of the castle, which was designed to house the women's rooms. It was not finished until 1892, after Ludwig's death. The original plan was to decorate the walls with statues of female saints, but these were never finished.

WELCOME TO THE CASTLE
The visitor enters the castle from the east, through two courtyards, the upper one shown here, and a lower one.

Statue of Liberty

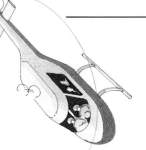

SEVEN SPIKES
The seven spikes of the crown represent the seven continents and seven seas over which enlightenment and liberty would spread.

CROWNING GLORY
The total height to the top of Liberty's crown is 306 ft. 8 in. (93.5 m). Of this total 152 ft. 2 in. (46.4 m) is accounted for by the statue itself, the rest by the pedestal she stands on.

FACIAL FEATURES
The statue's nose is 3 ft. 3 in. (1 m) long, her eyes are 2 ft. 6 in. (76 cm) across, and her mouth is 3 ft. 5 in. (90 cm) wide.

EYE LEVEL
As well as boat trips to Liberty Island, there are helicopter trips around the statue. At this level, the giant proportions and sculpted features of the statue can be seen clearly. The sculptor used his mother's strong facial features as a model for the statue.

TO CELEBRATE THE 100TH anniversary of American Independence, in 1884, the people of France presented the United States of America with an extraordinary gift as a token of friendship. It was an enormous statue (one of the largest ever built) of a woman stepping out of broken chains and holding the torch of freedom in her raised right hand. This figure of "Liberty enlightening the world" has come to be known simply as the Statue of Liberty. The story of how the statue was created is amazing in itself. It took 15 years (1870-1885) to build in Paris, with money raised by the French people through lotteries and dinner parties.

The statue was designed by the French sculptor Frederic Auguste Bartholdi. The inner iron structure was the work of Gustave Eiffel, the designer of the Eiffel Tower in Paris. Made of 32 tons (32.5 tonnes) of beaten copper from Norway, the statue was assembled first in Paris and then taken apart for shipping. The various parts of the statue traveled across the Atlantic, on a French ship, in 210 crates. When it arrived in New York, there was a 15-month delay because the pedestal (base), was not ready. The statue was finally unveiled on October 28, 1886, to the amazement of onlookers. Since then, it has stood as a reminder of the love of freedom of the two countries that built it.

The various sections of the statue were made in the studios of Gaget, Gauthier, & Co. in Paris, starting with the hand and torch. Copper was chosen because it was cheap, light, and flexible. Three hundred sheets of metal were beaten into shape over a wooden frame. This copper "skin" is no thicker than a small coin!

THE LIGHT OF LIBERTY
The lighting of Liberty's torch has caused quite a few problems over the years. Bartholdi thought the best method was to shine strong light onto the gilded flame from the torch platform. But this was rejected for fear it would dazzle people in ships in the harbor. Instead, a light was put inside the flame and holes were cut out so that light could shine through. This plan did not work and eventually Bartholdi's original idea was taken up. Today, the flame glows with the reflection of strong lights (which do not dazzle), just as he had intended.

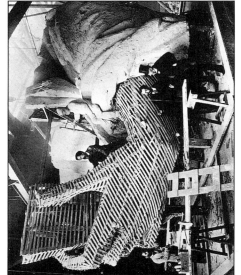

Tablet

The tablet carried by the statue is 24 ft (7.3 m) long and represents the tablets of law. Written on it, in Roman numerals, is "July 4, 1776," the day the United States declared its independence.

Materials

When copper is exposed to moist air it turns green. This process has given the statue its beautiful green color. However, the copper did suffer from corrosion. In the 1980s major restoration work was carried out to clean Liberty and give her further protection from aging and the elements.

In the base

The base is 152 ft. 2 in. (46.4 m) high. As well as an open stairway, a glass-walled elevator now takes visitors to the top of the pedestal. This is a good way to see the statue's inner workings. A museum at the base of the pedestal, which was opened in the 1960s, charts the history of the statue. Visitors can view the harbor from the high balcony.

Framework

Gustave Eiffel used the technology of bridge building when he designed the statue's inner framework. It is made up of wrought iron bars that cross each other diagonally to give maximum support to the frame.

When the wind blows

The thin outer structure or "skin" is attached to the inner frame supports by flat iron bars. These bars act much like springs, allowing the outer structure the flexibility to withstand winds of up to 125 miles (200 km) per hour.

Winding up

A metal spiral staircase winds up through the central tower, which is 110 feet (33.5 m) high. There are 171 steps to take the visitor up to the viewing gallery in the crown. At five different levels during the climb, visitors can step off onto rest platforms and view the inside of the statue.

Room with a view

Many visitors climb the steps to the viewing gallery in the head of the statue. From here they get a spectacular view of the harbor and the New York skyline. The viewing gallery has 25 windows and can hold up to 30 people at a time.

On a pedestal

To double the height of the statue, so that it towered over the entrance to New York Harbor, the Americans provided a stone and concrete platform (pedestal) for it to stand on. Building of the pedestal began in 1884 and was finished in 1886, 15 months after the statue had arrived from France.

Poem

On the base of the statue, there is a poem by the American poet Emma Lazarus, entitled "The New Colossus" (1883).

The Statue stands on Liberty Island (once called Bedloe's Island) in New York Harbor. For millions of immigrants who came to America to escape the poverty and persecution in their own countries, the statue was a symbol of freedom. Liberty was the first sight that met their eyes as they sailed into New York Harbor after their long journey.

VAN EETVELDE HOUSE

TO BUILD A HOUSE WHERE the doorknob exactly fits the shape of a hand and the staircase opens out at the bottom as if to welcome the visitor was the idea of the Belgian architect Victor Horta. At the end of the 19th century, Horta was one of the key figures in an artistic movement called Art Nouveau (New Art), which recognized the possibilities in each building material, whether it was wood, stone, iron, or glass, and used them in a new and exciting way. The four-story house he built for the Baron van Eetvelde in Brussels is unique. The Baron was an official in the Belgian government and needed a house in which to entertain. He wanted luxury as well as comfort, and in this sense the Van Eetvelde House is one of Horta's most successful buildings. He used wrought iron, which until then had been used mainly in industrial architecture, in dramatic combination with materials such as wood and stained glass. Horta designed everything himself, from the mosaic floors and mahogany ceilings to the radiators and door hinges.

One of the most interesting features of Horta's design, and one that is typical of the Art Nouveau style, is the whiplash curve. This is a wavelike, spiraling curve that turns back on itself. These curves can be found all over the Van Eetvelde House – in the patterns of the stained-glass windows, the banisters, the fireplaces, the designs on carpets and floors, and in the carved woodwork. The main rooms of the house were a sitting room at the front, and a dining room at the rear. They are connected by a glass-roofed, octagonal (eight-sided) hallway, which is the most unusual part of the house.

This view across the octagonal hall, from the salon (sitting room) to the stained-glass door of the dining room, shows Horta's love of glass. The Van Eetveldes entertained so much that the two main rooms – the salon and the dining room – had to be spacious. This so-called transparent design, which had become popular with the fashion for conservatories in the 19th century, helped Horta create a feeling of space inside his buildings.

PLAN OF VAN EETVELDE HOUSE

1 Salon
2 Entrance
3 Dining room
4 Octagonal hall
5 Master bedroom
6 Bedrooms
7 Kitchen
8 Coal cellar
9 Staircase

GLASS ROOF
Horta used glass wherever he could in the Van Eetvelde House, to let in as much light as possible. The inner hall is topped by a spectacular stained-glass skylight, supported by slim iron posts and frames. Light coming through the glass of the roof throws a warm glow down into the space below, high-lighting the rich materials used in the hall – marble, mosaic, brass, and wood, such as mahogany.

IRONWORK
Horta believed that if "modern" materials such as iron are used in a building, they should be used decoratively, and they should be seen. The wrought-iron posts and banisters around the octagonal hall show this idea clearly. No other material available at the time would have given Horta such a perfect opportunity to create his characteristic Art Nouveau curves.

THE OCTAGON
This room, in the center of the house on the first floor, served a similar purpose to a hotel lobby. It provided a pleasant, airy place for visitors to wait for their hosts. The walls of the room are made of stone and marble, and the floor is covered in mosaics. Glass doors lead from the octagon to the main reception rooms – the salon and the dining room.

STAIRCASE
The staircase winds around the octagonal hall. Both the stairs and the landings are decorated with a swirling orange curved pattern, which is repeated in the wrought-iron banisters.

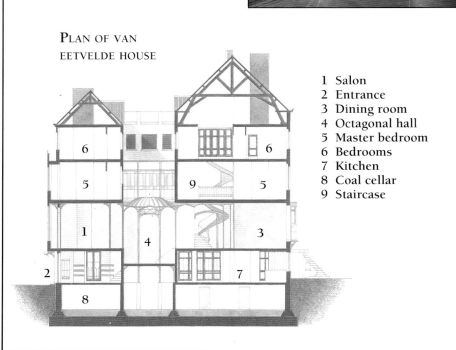

The outside of the Van Eetvelde
House in Brussels is not as richly
decorated as the inside. The only
hints of Art Nouveau are the
patterns on the mosaic panels
between floors, and on the
wrought-iron railings.

SEE-THROUGH

The salon, which occupies
the whole width of the
front of the house on the
first floor, can be glimpsed
through the glass screens
of the octagon. Horta
believed that a house
should be spacious and
adaptable, with good
communication between
rooms and plenty of light
and air getting into them.

FURNITURE

Horta often designed the
furniture for his buildings.
This five-level wooden table
in the octagonal hall was
made specially for the
Van Eetvelde House.

The metal posts
reflected in the salon
mirror carry the main
front wall of the house.

NOTRE-DAME-DU-HAUT

DURING WORLD WAR II, the old pilgrimage chapel of Notre-Dame-du-Haut at Ronchamp in eastern France was destroyed by gunfire. After the war, in 1950, the Swiss architect, Le Corbusier, was offered the job of designing a new chapel to replace it. In the next five years he created one of the most beautiful modern buildings, a place, as he said, "of silence, of prayer, of peace, of spiritual joy." The requirement was a small intimate building that would serve the local community as well as pilgrims who wanted to journey to it. The chapel stands on top of a hill at one end of the Vosges Mountains, near the Swiss border – a dramatic setting for the chapel. The sweeping roof and curving, whitewashed walls make up a shape that is more like a sculpture than a building. Each side looks completely different from the other, and the building has reminded people of a variety of things, from a crabshell to a dove, an airplane to a priest's hat. During the five years it took to build the chapel, Le Corbusier kept a notebook. In it he wrote of his feelings about the building: "The key is light, and light illuminates shapes, and shapes have emotional power."

The inside of the chapel is lit by many small stained-glass windows, set deep into the thick walls, like those in an ancient church or castle. These were designed and painted by Le Corbusier himself. A glass strip runs between the walls and the roof, letting a bright band of daylight into the chapel, and "floating" the roof free of the walls.

THE TALL TOWER
This is no ordinary tower. Its white, seamless, curved surfaces are unlike those of any other French church. They are more like the buildings Le Corbusier saw on trips to Greece.

Because of the difficulty of getting huge blocks of stone to the hilltop site, Le Corbusier used rubble from the ruined church to build the new chapel. He sprayed the walls with concrete to give a rough surface that could be whitewashed.

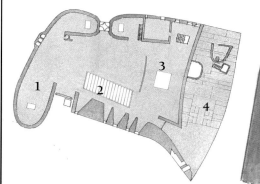

PLAN OF NOTRE-DAME-DU-HAUT
1 Side chapel in tower
2 Pews
3 Main chapel (holds 50 people)
4 Outdoor chapel

Main door

SIDE CHAPEL
Light filters through the windows in the back wall of one of the towers onto the small altar below.

MAIN CHAPEL
The inside of the chapel is simple and restful, with only a small number of pews and ornaments.

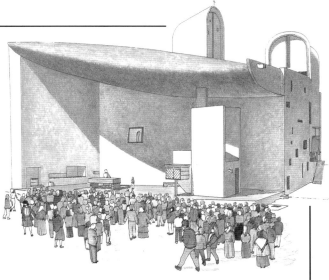

THE ROOF

Approaching the chapel from a distance, you may wonder how it is put together. Is the roof solid or hollow? How is it supported? In fact it is a hollow shell, and much of the support comes from the thick south wall of the building.

ROOF STRUCTURE

The hollow roof of the chapel is made of reinforced concrete, built over a framework of metal beams and bars. It is protected by several coats of waterproof material.

OUTSIDE SERVICES

At special festivals, when the congregation is too large for the indoor chapel, services are held outside. Le Corbusier provided an open-air chapel, with an altar and pulpit sheltered beneath the overhang of the roof. The curved wall behind helps project the priest's voice out to the congregation. The chapel, and the picturesque scenery of the Vosges, provides a fitting backdrop.

Le Corbusier was also a painter, who exhibited under his real name, Charles Edouard Jeanneret. The main door of the chapel is decorated with his painted panels, which show clouds, stars, an open hand, and a winding river. These pictures suggest a religion based on a love of the natural world.

Statue of the
Virgin Mary

BACK VIEW

A view of the back of the chapel shows how very different it looks as you walk around it. From this angle the roof is barely visible. The huge spout on the right drains water from the roof.

Outdoor chapel

THE GUGGENHEIM MUSEUM

AT THE CORNER OF FIFTH AVENUE in New York, facing Central Park, stands one of the most original buildings in the world – the Guggenheim Museum. It was commissioned in 1943 by Solomon R. Guggenheim, the American art collector, to house his collection of modern paintings. Frank Lloyd Wright, perhaps the most famous architect of the time (and one of the most unconventional), was chosen to design it. His love of unusual shapes led him to create a circular gallery formed by a spiral ramp that gets wider as it climbs upward. The planning took many years and the building was not completed until 1959. In the end Wright provided Guggenheim with just what he wanted: "a building that would be more than just an ordinary museum for hanging pictures." The idea was that visitors to the museum would enjoy the building as well as the exhibits. Wright designed an elevator to take people up to the top level, so that they could slowly wind their way down the ramp, looking at the paintings as they went. This way, visitors would not get too tired walking around the museum, and the paintings would be displayed in one sequence. The museum has recently been renovated and added to, but this has not altered the original design. Today the Guggenheim is one of the great landmarks of New York.

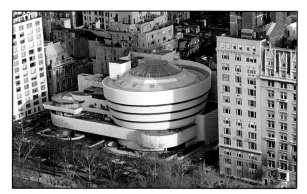

The concrete circles and spirals of the Guggenheim contrast sharply with the surrounding high-rise buildings of New York. When it was first built some people thought a flying saucer had landed.

A WORLD OF ITS OWN
Once inside the museum you seem to enter another world, far from the life of the city. There are no views out, so the visitor can concentrate on looking at the paintings.

SPIRALING RAMP
The descending ramp seems to float in midair, as if suspended in space. In reality, the ramp projects from 12 slabs, visible on the inside as partitions between groups of paintings. These cannot be seen from the outside, a fact that gives the building a timeless, almost magical quality.

SKYLIT DOME
The gallery roof is a glass dome that lets light into the center of the museum. People inside cannot see the sky through the frosted glass, another example of the way the building shuts out the outside world.

PLAN OF THE GUGGENHEIM
(AT GROUND LEVEL)
1 Main gallery
2 Grand gallery
3 Café
4 Ramp
5 Main elevator
6 Elevator
7 Entrance
8 Information desk

FLOATING IN SPACE
The light coming through the glass dome high-lights the space at the center of the museum. The ramp was meant to encircle the central space as if the building was spinning around it between heaven and earth.

A PROBLEM WITH LIGHTING
Wright's design included narrow bands of glass between the levels of the ramp. Unfortunately, instead of lighting the pictures naturally, the daylight shining through the glass dazzled people as they looked at the paintings. To solve this problem, extra artificial lights were installed.

PRIVATE VIEW
Supporting walls also act as partitions between groups of paintings, allowing the visitor to look at a few at a time.

PICTURE HANGING
The walls of the side galleries were angled backward to follow the lean of the ramp. Wright said that it was as if the paintings were still on the artist's easel, but the curve made it difficult to hang large pictures and so they were eventually hung vertically.

THE CENTRAL VOID
Standing at ground level and looking up at the spiraling galleries, visitors get a wonderful sense of space.

UNDERGROUND ROOMS
Tucked away in the basement, beneath the circular gallery, are a lecture theater, offices, archives, workrooms, and a café.

SYDNEY OPERA HOUSE

IF YOU SAIL INTO SYDNEY HARBOUR in Australia, you will see an extraordinary building, like a huge yacht, jutting into the cove. This is the Sydney Opera House, often called the eighth wonder of the world. It was built as the result of an international competition held in 1956 to find the best design for a new arts center. Many people entered the competition, but the most striking design, which stood out for its originality and imagination, was the one handed in by Jörn Utzon, a young Danish architect. The project was started in 1959 and the building was completed 14 years later, in 1973. The opening ceremony was a glittering occasion, with yachting races and a spectacular firework display. Although it was not built just for operas, people began to call the building the "Sydney Opera House," and the name stuck. It has four main halls (Opera Theater, Concert Hall, Playhouse, and Drama Theater), and includes a library, exhibition hall, and two main restaurants.

The Opera House stands on the headland of Bennelong Point, a peninsula named after the first English-speaking aborigine to live there. It is one of the finest sites in the world because it allows the building to be viewed from all angles, every one spectacular.

PLAN OF THE SYDNEY OPERA HOUSE
1 Opera stage   5 Stairs to foyers
2 Opera seating   6 Bennelong Restaurant
3 Concert stage   7 Concourse (under steps)
4 Concert seating 8 Main staircase

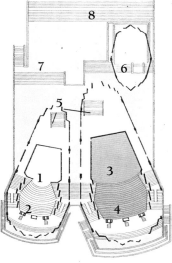

THE APPROACH
The visitor approaches the Opera House up a huge flight of steps, leading from the Botanic Gardens.

OPERA THEATER
The main hall on the left is for opera performances. It seats 1,547 people.

WALLS OF GLASS
People inside the building have a magnificent view of the harbor from the glass-fronted foyers.

Harbour Restaurant

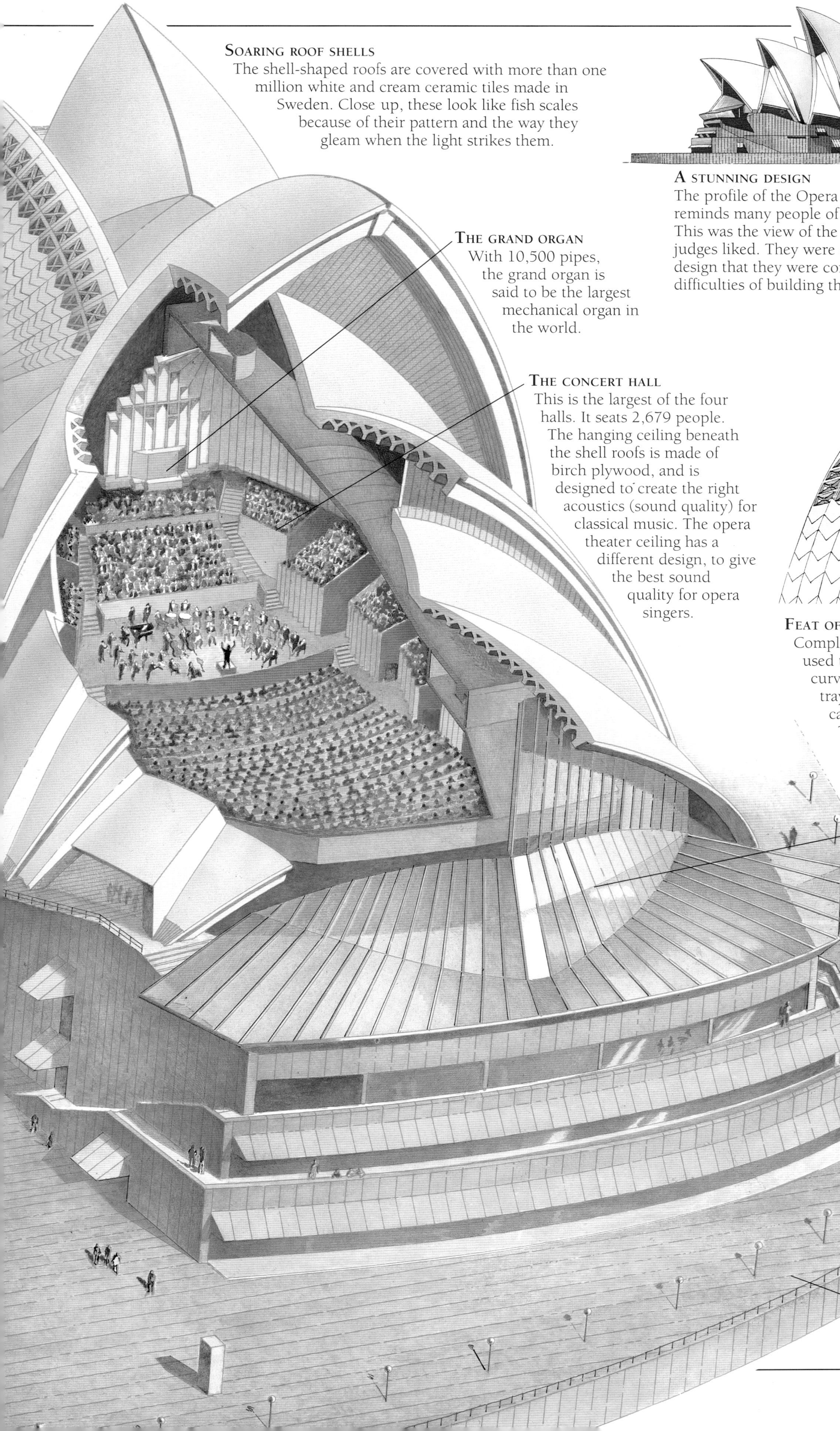

SOARING ROOF SHELLS
The shell-shaped roofs are covered with more than one million white and cream ceramic tiles made in Sweden. Close up, these look like fish scales because of their pattern and the way they gleam when the light strikes them.

A STUNNING DESIGN
The profile of the Opera House, with its overlapping roofs, reminds many people of the sails on boats in the harbor. This was the view of the building that the competition judges liked. They were so impressed with Utzon's brilliant design that they were convinced they could overcome the difficulties of building the Opera House.

THE GRAND ORGAN
With 10,500 pipes, the grand organ is said to be the largest mechanical organ in the world.

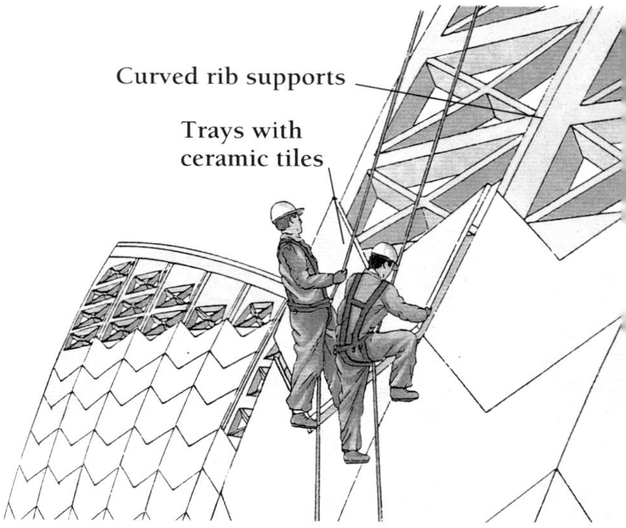

THE CONCERT HALL
This is the largest of the four halls. It seats 2,679 people. The hanging ceiling beneath the shell roofs is made of birch plywood, and is designed to create the right acoustics (sound quality) for classical music. The opera theater ceiling has a different design, to give the best sound quality for opera singers.

Curved rib supports

Trays with ceramic tiles

FEAT OF ENGINEERING
Complicated computer calculations were used to work out how to build the curved roof structure, and how to place trays of tiles onto it. The work itself called for a certain amount of daring. The top of the tallest shell roof is 221 feet (67.4 m) above sea level – as tall as a 22-story office block.

BRILLIANT USE OF GLASS
The glass beneath the shell roofs was made in France and is unique to the Opera House. Two layers were used to protect against the heat, and block the sound of ships' sirens in the harbor.

"Strollers" terrace around building

THE TORONTO SKYDOME

IT IS BIGGER THAN THE COLOSSEUM, as tall as a 31-story building, and seats 50,000 people. SkyDome would be impressive even if it was an ordinary stadium, but it is much more. It is the world's first large stadium with a moving roof that can open and close to provide a vast open-air arena in summer, and a domed, weatherproof stadium for autumn and winter events. SkyDome stands on old railway lands in the center of Toronto, one of Canada's major cities. It is the home ground for the Blue Jays baseball team and the Toronto Argonauts football team. With a daily workforce of 2,300 people, it took only three years to build the stadium, which opened in the summer of 1989. The building is a fantastic feat of engineering: in addition to the moving roof, there are thousands of seats that can be rotated in sections to suit the occasion. For smaller events SkyDome can be divided in half to make "SkyBowl," which has excellent sound quality for concerts. For special shows (and with the help of 89 huge curtains!) this area can be transformed into "SkyTent," which seats up to 27,000 people. Such adaptability means that the stadium can host almost any kind of entertainment, from football and baseball games to rock concerts and circuses. Other events have included boxing matches, motorcycle races, horse shows, and exhibitions.

SkyDome is in Toronto city center, close to the famous CN (Canadian National) Tower. It was built in a central location so that it would be easy to get to on public transport. A covered walkway, called SkyWalk, leads from Union Station to SkyDome and the CN Tower.

LIGHTING THE SKYDOME
The stadium has an impressive display of field lights: more than 700 units and 9,000 lightbulbs. Each unit gives a dazzling 2,000-watt illumination. Altogether, 120 miles (193 km) of electrical cable has been laid in SkyDome.

CAPACITY
Seating is on four levels. The lowest is the Esplanade (23,000 seats); the middle tier, SkyClub, has 13,500 seats; the next level, SkyBox, has 161 private boxes; the top tier, SkyDeck, seats 20,000.

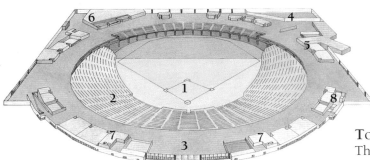

PLAN OF THE TORONTO SKYDOME

1 Arena
2 Seating area
3 Ceremonial entrance
4 Entertainment center
5 Restaurant
6 Sports bar
7 Elevators
8 Stairs

THE FIELD

SkyDome's field covers an area of 8 acres (3.25 hectares), big enough to fit eight jumbo jets, more than 1,000 elephants, or the Roman Colosseum!

TO TOP IT ALL

The roof weighs an incredible 11,000 tons (tonnes). It is made up of four panels, three of which are mobile. The fourth is fixed and provides shade for the video scoreboard. The panels are made of steel and covered with a PVC membrane. They are able to rotate by means of 54 drive mechanisms, called "bogies," which run on steel tracks.

OPENING SEQUENCE

SkyDome's roof takes 20 minutes to open or close. When the roof is open, 100 percent of the field and 91 percent of the seats are uncovered. The sequence (left) shows how the roof panels move back to an open position.

1 Roof fully closed.
2 Panels two and three start to slide back (retract) on parallel tracks.
3 Panels two and three retracted; panel one turns on a circular track.
4 Panel one completes 180° turn.
5 Roof open.

JUMBOTRON

SkyDome has the world's largest video screen, a Sony Jumbotron. Measuring 100 ft x 33 ft (33.6 m x 10 m), it is a scoreboard as well as a screen for viewing action reruns and close-ups of performers.

ALL ZIPPED UP

The field is covered with 106 rolls of Astroturf, joined together with 8 miles (12.8 km) of zippers – enough for 50,000 pairs of jeans! It takes 10-12 hours to zip or unzip the turf to convert the field from one sport to another.

INDEX

Dorling Kindersley would like to thank:
Peter Radcliffe and Christopher Gillingwater for additional design help; Richard Ward and Barrie Thorpe for additional illustrations; Janet Abbot for additional research; Tracey Moore and Kate Raworth for providing reference; George Hart of the British Museum for his help and advice.

Picture Research Catherine O'Rourke

Picture Credits
Bastin & Evrard 38cl, 39tr, 39br
Bridgeman Art Library; Private Collection **Shah Jehan and his wife Mumtaz Mahal** 24tr; by courtesy of the Board of Trustees of Victoria and Albert Museum **Portrait of Shah Jehan** 24tl; Johnny van Haeften Gallery, London **Louis XIV** by Jean Petitot the Elder 1607-1691 26tl;
Chris Fairclough Colour Library 18tr
Werner Forman 14cl, 22tr, 22cl
La Goelette, Paris 40tr

Sonia Halliday Photographs: Jane Taylor 12cl, 19tr
Robert Harding Picture Library 16cl, 44cl;
Rainbird **Versailles** by Pierre Patel 26 tr; **The Dream King** by Richard Wenig 1880 34tr
Michael Holford 7tl, 10cl
Hulton-Deutsch 36br
Hutchinson Picture Library 8cl, 18tl, 24bl;
Felix Greene 19bc
Image Bank: Alan Becker 46tr;
P & G Bowater 30tr; David W. Hamilton 16bl;

Tadao Kimura 26cl; N. Mareschal 29cr;
Marvin E. Newmann 15tl; Andrea Pistoli 37bl;
J. Ramey 16tr; Santi Visalli 42cl
A. F. Kersting 33tr
Scala 20bl
Andrew C Smith 41cr
Zefa 35tr; Bahnsen 40bl